Mechanical Engineering
Craft Studies
Part 1

A. Greer C Eng, MRAeS
Senior lecturer in Mechanical Engineering
Gloucester City College of Technology

W.H. Howell C Eng, MIProdE, MIQA, MISM
Lecturer in Mechanical Engineering
Gloucester City College of Technology

Edward Arnold

Preface

This book contains all the work required for the Mechanical Engineering Craft Studies Course (CGLI subject number 200). It has been found convenient to divide the book up into four sections—Calculations, Craft science, Engineering drawing and Craft theory. The SI system of units has been used throughout.

Since the course has been designed for craft level students who are undergoing the EITB recommended first year training, we have thought it unnecessary to include practical exercises. However, we have included descriptions of hand and machine tools and much other material which will help instructors and teachers involved in practical training. For this reason the book may be used both in the Technical Colleges and in the training workshops.

In the Calculation and Science sections it has been thought advisable to start from the beginning because many students find all types of calculations extremely difficult. It should be noted that logarithms and the slide rule are optional parts of the syllabus, but the more able students will benefit from these topics. For the less able students it may be as well to leave these topics until the second year of the course. The Drawing section is comprehensive, the aim being to enable the student to read engineering drawings with a fair degree of proficiency at the end of his first year's training. In the Craft theory section the aim has been to relate practice and theory, so that the student knows not only how to do the job but also knows the reason why he does it.

At the end of each chapter exercises relevant to the material in that chapter have been included. These questions are of the traditional type. The examining boards have decided to rely, to a large extent, on objective-type questions in their examinations. We have, therefore, compiled a number of Revision Questions which are of the objective type. However, we feel that traditional-type questions play an important part because students must be taught to express themselves clearly and as briefly as possible.

We wish to thank many colleagues and friends who have assisted us with their helpful advice and criticism; the firms who have allowed us to use their photographs and diagrams and the publishers who have given us a considerable amount of help in preparing the manuscript. Finally, our thanks are due to the City and Guilds of London Institute and other examining bodies for granting permission to use examples taken from their examination papers.

A.G.
W.H.H.

Acknowledgement

The following figures are based on illustrations appearing in *Basic Engineering Craft Course: Workshop Theory* by R. L. Timings published by Longman Group Limited.

Figs. 4.36, 4.40, 4.42, 4.47, 4.50, 4.53, 4.147, 4.175, 4.183, 4.184, 4.199, 4.212 and 4.235.

The Authors take this opportunity of gratefully acknowledging permission to use these illustrations.

Contents

Part 1   Calculations

1.1   Fractions

Vulgar fractions. The circle in Fig. 1.1 is divided into a number of equal parts, in this case eight. Each part is called one-eighth of the circle and is written as $\frac{1}{8}$. The number 8 below the line shows how many equal parts there are and it is called the *denominator*. The number above the line shows how many parts are taken and it is called the *numerator*. If seven of the eight parts are taken then we have $\frac{7}{8}$ of the circle.

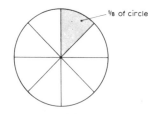
⅛ of circle

Fig. 1.1

The value of a fraction is unchanged if we multiply or divide both its numerator and denominator by the same amount.

$\frac{4}{5} = \frac{12}{15}$ (by multiplying numerator and denominator by 3)

$\frac{2}{3} = \frac{14}{21}$ (by multiplying numerator and denominator by 7)

$\frac{12}{32} = \frac{3}{8}$ (by dividing numerator and denominator by 4)

$\frac{8}{64} = \frac{1}{8}$ (by dividing numerator and denominator by 8)

Sometimes we can divide the numerator and denominator of a fraction by the same number several times. This is called 'reducing the fraction to its lowest terms'. For example,

$\frac{168}{378} = \frac{84}{189}$ (by dividing numerator and denominator by 2)

$= \frac{28}{63}$ (by dividing numerator and denominator by 3)

$= \frac{4}{9}$ (by dividing numerator and denominator by 7)

Lowest common denominator. The L.C.M. (least common multiple) of a set of separate numbers is the smallest number into which each of the given numbers will divide. Thus, the L.C.M. of 4, 5 and 10 is 20. That is, 20 is the smallest number into which 4, 5 and 10 will divide. The L.C.M. of 2, 6 and 8 is 24.

Example. Arrange the following fractions in order of size, beginning with the smallest: $\frac{3}{4}$, $\frac{5}{8}$, $\frac{7}{10}$ and $\frac{11}{20}$.

First find the L.C.M. of the denominators 4, 8, 10 and 20. The L.C.M. is 40. Now express each of the given fractions with denominator 40.

$$\frac{3}{4} = \frac{3 \times 10}{4 \times 10} = \frac{30}{40}, \quad \frac{5}{8} = \frac{5 \times 5}{8 \times 5} = \frac{25}{40},$$

$$\frac{7}{10} = \frac{7 \times 4}{10 \times 4} = \frac{28}{40}, \quad \frac{11}{20} = \frac{11 \times 2}{20 \times 2} = \frac{22}{40}.$$

The order is $\frac{22}{40}$, $\frac{25}{40}$, $\frac{28}{40}$ and $\frac{30}{40}$ or $\underline{\frac{11}{20}}$, $\underline{\frac{5}{8}}$, $\underline{\frac{7}{10}}$ and $\underline{\frac{3}{4}}$.

Exercise 1.1

Find the L.C.M. of:
1. 4 and 6
2. 12 and 10
3. 2, 6 and 12
4. 3, 4 and 5
5. 10, 8 and 2
6. 3, 8 and 6
7. 20 and 25
8. 32 and 20

Arrange in order of size, beginning with the smallest, the following groups of fractions:

9. $\frac{1}{2}$, $\frac{5}{6}$, $\frac{2}{3}$, $\frac{7}{12}$
10. $\frac{9}{10}$, $\frac{3}{4}$, $\frac{6}{7}$, $\frac{7}{8}$
11. $\frac{13}{16}$, $\frac{11}{20}$, $\frac{7}{10}$, $\frac{3}{5}$
12. $\frac{3}{4}$, $\frac{5}{8}$, $\frac{2}{3}$, $\frac{13}{20}$

Types of fractions. If the numerator of a fraction is *less* than its denominator, the fraction is called a *proper fraction*. Thus, $\frac{2}{3}$, $\frac{5}{8}$, $\frac{7}{16}$ and $\frac{3}{4}$ are all proper fractions.

If the numerator of a fraction is *greater* than its denominator, then the fraction is called an *improper fraction*. Thus $\frac{11}{8}$, $\frac{9}{7}$, $\frac{5}{2}$ and $\frac{7}{7}$ are all improper fractions. All improper fractions have a value greater than 1.

Every improper fraction can be expressed as a whole number and a proper fraction. These are sometimes called *mixed numbers*. Thus $1\frac{3}{8}$, $2\frac{1}{2}$ and $3\frac{2}{5}$ are all mixed numbers.

In order to convert an improper fraction to a mixed number it must be remembered that

$$\frac{Numerator}{Denominator} = Numerator \div Denominator$$

$\frac{15}{7} = 15 \div 7 = 2$ and remainder $1 = 2\frac{1}{7}$

We can therefore convert an improper fraction to a mixed number by dividing the denominator into the numerator. It will be noticed, from the example above, that the remainder becomes the numerator in the fractional part of the mixed number.

Examples

1. Express as mixed numbers (a) $\frac{15}{8}$ (b) $\frac{14}{7}$ (c) $\frac{9}{2}$

(a) $\frac{15}{8} = 1\frac{7}{8}$ (since $15 \div 8 = 1$ and remainder 7)

(b) $\frac{14}{7} = 2$ (since $14 \div 7 = 2$ exactly)

(c) $\frac{9}{2} = 4\frac{1}{2}$ (since $9 \div 2 = 4$ and remainder 1)

2. Express as improper fractions (a) $3\frac{3}{4}$ (b) $1\frac{5}{8}$

(a) $3\frac{3}{4} = \frac{3 \times 4}{4} + \frac{3}{4} = \frac{12}{4} + \frac{3}{4} = \frac{15}{4}$

(b) $1\frac{5}{8} = \frac{1 \times 8}{8} + \frac{5}{8} = \frac{8}{8} + \frac{5}{8} = \frac{13}{8}$

Exercise 1.2

Express each of the following as a mixed number:

1. $\frac{5}{2}$ 2. $\frac{8}{3}$ 3. $\frac{22}{7}$ 4. $\frac{11}{10}$
5. $\frac{19}{8}$ 6. $\frac{15}{12}$ 7. $\frac{24}{9}$ 8. $\frac{41}{20}$

Express each of the following as an improper fraction:

9. $2\frac{1}{10}$ 10. $3\frac{1}{4}$ 11. $1\frac{7}{32}$ 12. $2\frac{3}{64}$
13. $3\frac{2}{7}$ 14. $4\frac{7}{20}$ 15. $8\frac{2}{3}$ 16. $7\frac{3}{8}$

Addition of fractions. In the last example we added together two fractions. As each had the same denominator we merely added together the two numerators. The fraction we obtained had this sum as the numerator with the original denominator. This example points the way in which fractions are added.

Example. Find the sum of $\frac{3}{5}$ and $\frac{1}{3}$.

First find the L.C.M. of the denominators 5 and 3. This is 15. Now express $\frac{3}{5}$ and $\frac{1}{3}$ with a denominator of 15.

$$\frac{3}{5} = \frac{3 \times 3}{5 \times 3} = \frac{9}{15}$$

$$\frac{1}{3} = \frac{1 \times 5}{3 \times 5} = \frac{5}{15}$$

$$\frac{3}{5} + \frac{1}{3} = \frac{9}{15} + \frac{5}{15} = \frac{9+5}{15} = \frac{14}{15}$$

A better way is to set out the work as follows:

$$\frac{3}{5} + \frac{1}{3} = \frac{3 \times 3 + 1 \times 5}{15} = \frac{9+5}{15} = \frac{14}{15}$$

Example. Simplify $\frac{3}{4} + \frac{2}{3} + \frac{7}{10}$.

The L.C.M. of the denominators 4, 3 and 10 is 60.

$$\therefore \qquad \frac{3}{4} + \frac{2}{3} + \frac{7}{10} = \frac{15 \times 3 + 20 \times 2 + 6 \times 7}{60}$$

$$= \frac{45 + 40 + 42}{60} = \frac{127}{60} = 2\frac{7}{60}$$

Example. Simplify $4\frac{1}{2} + 1\frac{2}{3} + 3\frac{2}{5}$.

The whole numbers are added together and then the fractional parts are dealt with. The L.C.M. of the denominators 2, 3 and 5 is 30.

$$\therefore \qquad 4\frac{1}{2} + 1\frac{2}{3} + 3\frac{2}{5} = 8\frac{15 \times 1 + 10 \times 2 + 6 \times 2}{30}$$

$$= 8\frac{15 + 20 + 12}{30}$$

$$= 8\frac{47}{30}$$

$$= 8 + \frac{47}{30} = 8 + 1\frac{17}{30} = 9\frac{17}{30}$$

Exercise 1.3

Simplify:

1. $\frac{1}{2} + \frac{1}{3}$ 2. $\frac{3}{5} + \frac{7}{10}$ 3. $\frac{3}{4} + \frac{3}{8}$
4. $\frac{2}{5} + \frac{1}{4}$ 5. $\frac{1}{2} + \frac{1}{4} + \frac{5}{8}$ 6. $\frac{3}{10} + \frac{7}{100} + \frac{31}{1000}$
7. $1\frac{3}{8} + 2\frac{5}{16}$ 8. $6\frac{2}{3} + 3\frac{2}{5}$ 9. $3\frac{5}{16} + 2\frac{3}{64}$
10. $4\frac{1}{2} + 3\frac{1}{6} + 2\frac{1}{3}$ 11. $5\frac{3}{8} + 4\frac{3}{4} + \frac{5}{8} + \frac{7}{16}$ 12. $5\frac{2}{3} + \frac{3}{5} + \frac{7}{10} + 2\frac{1}{2}$

Subtraction of fractions.

Example. Simplify $\frac{3}{8} - \frac{1}{5}$.

The L.C.M. of the denominator is 40.

$$\therefore \qquad \frac{3}{8} - \frac{1}{5} = \frac{5 \times 3 - 8 \times 1}{40} = \frac{15 - 8}{40} = \frac{7}{40}$$

Where mixed numbers have to be subtracted, the easiest way is to convert the mixed numbers to improper fractions and proceed in the way shown in the last example.

Example. Simplify $2\frac{7}{10} - 1\frac{1}{4}$.

$$2\frac{7}{10} - 1\frac{1}{4} = \frac{27}{10} - \frac{5}{4} = \frac{2 \times 27 - 5 \times 5}{20} = \frac{54 - 25}{20}$$

$$= \frac{29}{20} = 1\frac{9}{20}$$

Example

$$2\frac{2}{5} - 1\frac{5}{7} = \frac{12}{5} - \frac{12}{7} = \frac{7 \times 12 - 5 \times 12}{35} = \frac{84 - 60}{35} = \frac{24}{35}$$

Exercise 1.4

1. $\frac{1}{2} - \frac{1}{3}$ 2. $\frac{1}{3} - \frac{1}{7}$ 3. $\frac{2}{3} - \frac{1}{2}$
4. $\frac{7}{10} - \frac{3}{10}$ 5. $\frac{7}{8} - \frac{5}{6}$ 6. $3\frac{1}{2} - 1\frac{3}{8}$
7. $2 - \frac{7}{8}$ 8. $6 - 3\frac{2}{5}$ 9. $4\frac{3}{8} - 2\frac{7}{10}$
10. $1\frac{5}{16} - \frac{4}{5}$ 11. $4\frac{5}{32} - 2\frac{17}{20}$

Example. Simplify $4\frac{3}{8} - 2\frac{1}{4} + \frac{1}{5}$.

$$4\frac{3}{8} - 2\frac{1}{4} + \frac{1}{5} = \frac{35}{8} - \frac{9}{4} + \frac{1}{5}$$

$$= \frac{5 \times 35 - 10 \times 9 + 8 \times 1}{40}$$

$$= \frac{93}{40} = 2\frac{13}{40}$$

Exercise 1.5

Simplify:

1. $3\frac{1}{2} + 2\frac{1}{4} - 3\frac{3}{8}$
2. $5\frac{1}{10} - 2\frac{1}{2} - 1\frac{1}{4}$
3. $5\frac{3}{8} - 4\frac{1}{2} + 7$
4. $5\frac{1}{2} - 2\frac{1}{4} + 2\frac{1}{12} - 4\frac{3}{4}$
5. $1\frac{1}{16} - 2\frac{1}{2} - 3\frac{1}{4} + 5\frac{5}{8}$
6. $3 + 2\frac{1}{4} - 4\frac{3}{4}$
7. $2\frac{3}{16} - 2\frac{3}{10} + \frac{5}{8}$
8. $12\frac{1}{4} - 6\frac{1}{8} - 3\frac{3}{64}$
9. $3\frac{9}{20} - 1\frac{5}{8} + \frac{1}{16}$
10. $2\frac{1}{25} + 3\frac{1}{5} - 2\frac{7}{10} - \frac{3}{20}$

Multiplication of fractions. When multiplying together two or more fractions we first multiply all the numerators together and then we multiply all the denominators together. Always convert mixed numbers into improper fractions.

$$\frac{5}{8} \times \frac{3}{7} = \frac{5 \times 3}{8 \times 7} = \frac{15}{56}$$

$$\tfrac{2}{5} \times 3\tfrac{2}{3} = \frac{2}{5} \times \frac{11}{3} = \frac{2 \times 11}{5 \times 3} = \frac{22}{15} = 1\tfrac{7}{15}$$

$$1\tfrac{3}{8} \times 2\tfrac{1}{4} = \frac{11}{8} + \frac{9}{4} = \frac{11 \times 9}{8 \times 4} = \frac{99}{32} = 3\tfrac{3}{32}$$

Example. Simplify $\tfrac{2}{3} \times 1\tfrac{7}{8}$.

$$\tfrac{2}{3} \times 1\tfrac{7}{8} = \tfrac{2}{3} \times \tfrac{15}{8} = \tfrac{30}{24} = \tfrac{5}{4} = 1\tfrac{1}{4}$$

The step of reducing $\tfrac{30}{24}$ to its lowest terms has been done by dividing 6 into both numerator and denominator. An easier method of doing this is the method of cancellation.

$$\overset{1}{\underset{1}{\frac{2}{3}}} \times \overset{5}{\underset{4}{\frac{15}{8}}} = \frac{1 \times 5}{1 \times 4} = \frac{5}{4} = 1\tfrac{1}{4}$$

We have divided 2 into 2 (a numerator) and 8 (a denominator) and 3 into 15 (a numerator) and 3 (a denominator). You will notice that we have divided the numerators and denominators by the same amount. Notice that we can only cancel between a numerator and a denominator.

Example. Simplify $\tfrac{15}{16} \times \tfrac{14}{27}$.

$$\overset{5}{\underset{8}{\frac{15}{16}}} \times \overset{7}{\underset{9}{\frac{14}{27}}} = \frac{5 \times 7}{8 \times 9} = \tfrac{35}{72}$$

Example. Simplify $\tfrac{16}{25} \times \tfrac{7}{8} \times 8\tfrac{3}{4}$.

$$\overset{\overset{1}{2}}{\underset{5}{\frac{16}{25}}} \times \overset{7}{\underset{1}{\frac{7}{8}}} \times \overset{7}{\underset{2}{\frac{35}{4}}} = \frac{1 \times 7 \times 7}{5 \times 1 \times 2} = \frac{49}{10} = 4\tfrac{9}{10}$$

Exercise 1.6
Simplify:

1. $\tfrac{3}{4} \times 1\tfrac{7}{9}$
2. $5\tfrac{1}{5} \times \tfrac{10}{13}$
3. $1\tfrac{5}{8} \times \tfrac{7}{26}$
4. $1\tfrac{1}{2} \times \tfrac{2}{5} \times 2\tfrac{1}{2}$
5. $\tfrac{1}{2} \times \tfrac{4}{7} \times \tfrac{3}{4}$
6. $\tfrac{5}{8} \times \tfrac{7}{10} \times \tfrac{2}{21}$
7. $2 \times 1\tfrac{1}{2} \times 1\tfrac{1}{3}$
8. $3\tfrac{3}{4} \times 1\tfrac{3}{5} \times 1\tfrac{1}{8}$
9. $\tfrac{3}{10} \times \tfrac{20}{9} \times 1\tfrac{1}{8}$
10. $\tfrac{15}{32} \times \tfrac{8}{11} \times 24\tfrac{1}{5}$

Division of fractions. When dividing fractions invert the division and multiply.

$$\tfrac{3}{5} \div \tfrac{2}{7} = \tfrac{3}{5} \times \tfrac{7}{2} = \tfrac{21}{10} = 2\tfrac{1}{10}$$

$$1\tfrac{4}{5} \div 2\tfrac{1}{3} = \tfrac{9}{5} \div \tfrac{7}{3} = \tfrac{9}{5} \times \tfrac{3}{7} = \tfrac{27}{35}$$

Exercise 1.7
Simplify:

1. $\tfrac{4}{5} \div \tfrac{5}{9}$
2. $1 \div \tfrac{1}{4}$
3. $1 \div 1\tfrac{1}{2}$
4. $\tfrac{5}{8} \div \tfrac{15}{32}$
5. $3\tfrac{3}{4} \div 2\tfrac{1}{2}$
6. $2\tfrac{1}{2} \div 3\tfrac{3}{4}$
7. $2\tfrac{1}{10} \div \tfrac{3}{5}$
8. $3\tfrac{1}{15} \div 2\tfrac{5}{9}$
9. $5 \div 5\tfrac{1}{5}$
10. $\tfrac{7}{24} \div \tfrac{5}{8}$

1.2 Decimals

The decimal system. The decimal system is an extension of our ordinary number system. When we write the number 777 we mean $700+70+7$. Reading from left to right each figure 7 is ten times the value of the next one.

We now have to decide how to deal with fractional quantities, that is, quantities whose value is less than one. If we regard 777·777 as meaning $700+70+7+\frac{7}{10}+\frac{7}{100}+\frac{7}{1000}$, then the dot, called the *decimal point*, separates the whole numbers from the fractional parts. Notice that with the fractional or decimal part ·777, each figure 7 is ten times the value of the following one, reading from left to right. That is, $\frac{7}{10}$ is ten times as great as $\frac{7}{100}$, and $\frac{7}{100}$ is ten times as great as $\frac{7}{1000}$.

Decimals, then, are fractions in which the denominators are 10, 100, 1000 and so on, according to the position of the figure after the decimal point.

If we have to write 6 hundreds and 7 units we write 607; the 0 keeps the place for the missing tens. In the same way if we want to write $\frac{3}{10}+\frac{5}{1000}$ we write ·305; the 0 keeps the place for the missing hundredths.

Therefore, $\frac{5}{100}+\frac{3}{1000}$ would be written ·053; the 0 keeps the place for the missing tenths. Notice that if there are no whole numbers it is usual to insert a 0; that is 0·35 instead of ·35.

Exercise 1.8
Read off as decimals:
1. $\frac{3}{10}$
2. $\frac{5}{10}+\frac{7}{100}$
3. $\frac{2}{10}+\frac{4}{100}+\frac{3}{1000}$
4. $\frac{8}{1000}$
5. $\frac{3}{100}$
6. $\frac{1}{100}+\frac{7}{1000}$
7. $8+\frac{5}{100}$
8. $4+\frac{2}{100}+\frac{5}{10000}$
9. $50+\frac{4}{1000}$
10. $\frac{1}{10}+\frac{4}{100}+\frac{5}{10000}$

Read off fractions with denominators 10, 100, 10000, etc.
11. 0·2   12. 3·1   13. 2·46
14. 423·15   15. 0·003   16. 0·025
17. 300·026   18. 42·2385   19. 0·003 29
20. 0·0001

Addition and subtraction of decimals. Adding or subtracting decimals is done in exactly the same way as for whole numbers. Care must be taken, however, that the decimal points are all written under one another. This ensures that all the figures of similar place value fall in the same column.

Example. Add 11·27+3·265+0·023.

$$\begin{array}{r} 11\cdot27 \\ 3\cdot265 \\ \underline{0\cdot023} \\ \underline{14\cdot558} \end{array}$$

Example. Subtract 9·237 from 11·185.

$$\begin{array}{r} 11\cdot185 \\ \underline{9\cdot237} \\ \underline{1\cdot948} \end{array}$$

Exercise 1.9
Give the value of:
1. 1·375+0·625
2. 3·25+1·125
3. 2·196+1·735+8·359
4. 30·167+0·003+19·3
5. 27·318+0·143+18+1·328
6. 11·33−2·16
7. 18·216−3·195
8. 1·289−0·916
9. 0·875−0·053
10. 5·37−3·495

Multiplication and division of decimals. One of the advantages of decimals is the ease with which they may be multiplied or divided by 10, 100, 1000 and so on.

Example. Evaluate 1·3×10.
$$\begin{aligned} 1\cdot3\times10 &= 1\times10+\cdot3\times10 \\ &= 10+\tfrac{3}{10}\times10 \\ &= 10+3 \\ &= \underline{13} \end{aligned}$$

Example. Evaluate 196·253×10.
$$\begin{aligned} 196\cdot253\times10 &= 196\times10+\cdot2\times10+\cdot05\times10+\cdot003\times10 \\ &= 1960+\tfrac{2}{10}\times10+\tfrac{5}{100}\times10+\tfrac{3}{1000}\times10 \\ &= 1960+2+\tfrac{5}{10}+\tfrac{3}{100} \\ &= \underline{1962\cdot53} \end{aligned}$$

In both the above examples, you will notice that the figures have not been changed by the multiplication; only the *positions* of the figures have been changed. Thus, in the first example, $1\cdot3\times10 = 13$, that is the decimal point has been moved one place to the right. In the last example $196\cdot253\times10 = 1962\cdot53$; again the decimal point has been moved one place to the right.

To multiply by 10, then, is the same as shifting the decimal point one place to the right. In the same way to multiply by 100 means shifting the decimal point two places to the right and so on.

$$15\cdot128\times100 = 1512\cdot8;$$

decimal point moved 2 places to right.

$$0\cdot0365\times1000 = 36\cdot5;$$

decimal point moved 3 places to right.

$$0\cdot2\times100 = 20;$$

decimal point moved 2 places to right.

Exercise 1.10

Multiply each of the following by 10, 100, 1000.

1. $3\cdot1$ 2. $1\cdot32$ 3. $0\cdot081$ 4. $0\cdot25$

Give the value of:

5. $0\cdot1385\times100$ 6. $2\cdot7\times1000$

7. $170\cdot05\times10$ 8. $0\cdot3124\times10\,000$

9. $0\cdot007\times1000$ 10. $5\cdot132\times100$

In dividing by 10, the decimal point is moved one place to the left, by 100, two places to the left and so on. Thus,

$$102\cdot1 \div10 = 10\cdot21;$$

decimal point moved one place to left.

$$9\cdot16\div100 = 0\cdot0916;$$

decimal point moved two places to left.

$$5 \div1000 = 0\cdot005;$$

decimal point moved three places to left.

From the above examples, you will notice the use that has been made of the 0's following the decimal point; they keep the places of the tenths, hundredths, etc., which are missing.

Exercise 1.11

Divide each of the following by 10, 100, 1000.

1. $2\cdot5$ 2. $59\cdot163$ 3. $0\cdot05$ 4. $510\cdot3$

Give the value of:

5. $5\cdot2 \div100$ 6. $1\cdot05\div1000$

7. $78\cdot03 \div100$ 8. $0\cdot03\div10$

9. $0\cdot0096\div100$ 10. $725\cdot61\div10\,000$

Example. Find the value of $43\cdot2\times2\cdot503$.

First disregard the decimal points and multiply 432×2503. Thus,

$$
\begin{array}{r}
432 \\
2503 \\
\hline
864 \\
21600 \\
1296 \\
\hline
1081296
\end{array}
$$

Now count up the total number of figures after the decimal points in both of the quantities to be multiplied (in this case 4). In the answer to the multiplication count this total number of figures from the right, and insert the decimal point. The answer is then 108·1296.

Exercise 1.12

Find the value of the following:

1. $23\cdot12\times30\cdot16$ 2. $0\cdot0135\times2\cdot598$

3. $0\cdot75\times0\cdot25$ 4. $3\cdot015\times1\cdot75$

5. $0\cdot031\times0\cdot023$

Example. Find the value of $15\cdot75\div2\cdot5$.

First convert the divisor ($2\cdot5$) into a whole number by multiplying it by 10. To compensate for this multiply the dividend ($15\cdot75$) by 10; $15\cdot75$ then becomes $157\cdot5$. Now proceed as in ordinary division.

$$
\begin{array}{r}
6\cdot3 \\
25\overline{)157\cdot5} \\
\underline{150} \\
7\,5 \\
\underline{7\,5} \\
\cdots
\end{array}
$$

······This line = 25×6

······5 brought down from above. Since 5 lies to right of decimal point, insert decimal point into answer.

Notice how the position of the decimal point in the answer was obtained. The 5 brought down lies to the *right* of the decimal point in the dividend. Before bringing this down put the decimal point in the answer, that is, immediately after the 6

The division in this case is exact and the answer is $6\cdot3$. Now let us see what happens when there is a remainder.

Example. Find the value of $15\cdot187\div3\cdot57$.

In the same way as before, we make $3\cdot57$ into a whole number by multiplying it by 100, so that it becomes 357. We must now multiply $15\cdot187$ by 100 giving $1518\cdot7$.

$$
\begin{array}{r}
4\cdot25406 \\
357\overline{)1518\cdot7} \\
\underline{1428} \\
907 \\
\underline{714} \\
1930 \\
\underline{1785} \\
1450 \\
\underline{1428} \\
2200 \\
\underline{2142} \\
58
\end{array}
$$

··········This line = 357×4

··········7 brought down from above. Since 7 lies to right of decimal point insert decimal point in answer.

Bring down 0 as all figures above have been used up.

····Bring down 0. Divisor goes 0. Put 0 in answer and bring down another 0.

The answer to 5 decimal places is 4·254 06. This is not correct because there is a remainder. The division can be continued in the way shown, to give as many decimal places as desired, or until there is no remainder.

It is important to realize what is meant by an answer given, say, to 3 decimal places. Reading the answer, the first three decimal places give this as $4\cdot254$. However, if an answer correct to 4 decimal places is required, it will be $4\cdot2541$. This is because the 5th decimal place is 5 or greater (in fact, it is 6). When this happens the 4th decimal place is increased by one.

Thus, $91\cdot7683 = 91\cdot8$ correct to 1 decimal place.

$= 91\cdot77$ correct to 2 decimal places.

$= 91\cdot768$ correct to 3 decimal places.

If an answer is required correct to three decimal places, the division should be continued to four decimal places, and the answer expressed correct to three places.

Example. Find the value of $1\cdot3875\div0\cdot041$ correct to 2 decimal places.

Moving the decimal points as before,

```
        33·841
  41)1387·5
     123
     157
     123
     345
     328
     170
     164
          60    The answer correct to 2
          41    decimal places is 33·84.
          19
```

Of course, answers to multiplication problems can also be given correct to a number of decimal places.

Exercise 1.13
Find the value of:
1. 18·76÷14·3 correct to 2 decimal places.
2. 0·0396÷2·51 correct to 3 decimal places.
3. 7·21÷0·038 correct to 2 decimal places.
4. 13·059÷3·18 correct to 4 decimal places.
5. 0·1382÷0·0032 correct to 1 decimal place.

Significant figures. Instead of using the number of decimal places to give the accuracy of an answer significant figures can be used. The number 39·38 is correct to 2 decimal places but it is also correct to 4 significant figures, since the number contains 4 figures. The rules regarding significant figures are as follows:
(1) If the first figure to be discarded is 5 or more the previous figure in the answer is increased by 1.
    8·1925 = 8·193 correct to 4 significant figures
           = 8·19 correct to 3 significant figures
           = 8·2   correct to 2 significant figures
(2) Zeros must be kept to show the position of the decimal point, or to indicate that the zero is a significant figure.
  24 392 = 24 390 correct to 4 significant figures
         = 24 400 correct to 3 significant figures
  0·0858 = 0·086 correct to 2 significant figures
        = 0·09   correct to 1 significant figure
 413·703 = 413·70 correct to 5 significant figures
         = 414   correct to 3 significant figures
  0·0583 = 0·058 correct to 3 decimal places
        = 0·06   correct to 2 decimal places

Exercise 1.14
Write down the following numbers correct to the number of significant figures stated.
1. 18·865 74 (a) to 6 (b) to 4 (c) to 2
2. 0·008 269 1 (a) to 4 (b) to 3 (c) to 2
3. 4·968 48 (a) to 5 (b) to 3 (c) to 1
4. 48·978 to 2
5. 38·803 to 4
6. 40·586 517 (a) to 5 (b) to 2

7. 3·149 98 (a) to 5 (b) to 4 (c) to 3
8. 12·203 8 to 4

Approximations. In any decimal calculation, the worst mistake you can make is to misplace the decimal point. To place it wrongly, even by one place, makes the answer ten times too large or ten times too small. From the examples you have done already, you will be aware that it is very easy to misplace the decimal point. Approximations are made in order to prevent misplacement of the decimal point.

Example. Calculate 3·16×2·58.
For a rough estimate we will take 3×3 = <u>9.</u>
  Notice that 3·16 is near to 3, whilst 2·58 is nearer to 3 than 2.
Performing the multiplication.

```
        3·16
        2·58
        632
       1580
       2528
      8·1528
```

The answer is therefore <u>8·1528</u> not 81·528 or 0·815 28. If either of these incorrect answers had been produced, the rough estimate shows that the answer must be 8·1528.

Example. Calculate 0·235×0·384.
For a rough estimate take 0·2×0·4 = <u>0·08.</u>
Performing the multiplication,

```
       0·235
       0·384
        705
       1880
        940
     0·090240
```

Notice that the rough estimate gives the correct position of the decimal point in each of the above examples.
  Rough estimates are of most importance for division, where the danger of misplacing the decimal point is greater.

Example. Calculate 5·25÷2·41.
For a rough estimate we can take 5÷2 = <u>2·5</u>

```
          2·178
   241)525
       482
       430
       241
      1890
      1687
      2030
      1928
       102
```

The answer is 2·18 correct to two decimal places. Although the rough estimate is some way from the correct answer, it still tells us that the answer is 2·18 and not 21·8 or 0·218.

Example. Calculate 0·43÷2·07 roughly.
For a rough estimate take 0·4÷2 = 0·2

Example. Calculate 0·0156÷0·032 roughly.
For a rough estimate take 0·015÷0·03
$$= 1·5÷3$$
$$= 0·5$$

Exercise 1.15
Obtain rough estimates for the following and compare your rough estimate with the answer correct to 3 decimal places.
1. 0·026×3·189   2. 0·35×0·046   3. 2·683×3·197
4. 3·187÷1·258   5. 0·325÷2·56   6. 3·188÷0·219
7. 0·529÷0·187   8. 0·326÷0·002   9. 0·598÷4·27
10. 8·192÷0·018

Fraction to decimal conversion. We found, when doing fractions, that the line separating the numerator and denominator of a fraction takes the place of a division sign. Thus,

$\frac{13}{40}$ is the same as 13÷40.

Therefore, to convert a fraction into a decimal we divide the denominator into the numerator.

Example. Convert the following into decimals (a) $\frac{29}{64}$ (b) $2\frac{9}{16}$.

$$\frac{29}{64} = 29÷64$$

```
        0·453125
   64)290
      256
      340
      320
      200
      192
       80
       64
      160
      128
      320
      320
      ...        ∴ 29/64 = 0·453125
```

(b) Where we have a mixed number we need only deal with the fractional portion.
Thus, to convert $2\frac{9}{16}$ to decimal, we need only deal with $\frac{9}{16}$.

$$\frac{9}{16} = 9÷16$$

```
      0·5625
  16)90
     80
    100
     96
     40
     32
     80
     80
     ..      ∴ 9/16 = 0·5625 and 2 9/16 = 2·5625
```

Example. Convert $\frac{1}{3}$ to decimals.

$$\frac{1}{3} = 1÷3$$

```
     0·333
  3)1·0
    9
   10
    9
   10
    9
    1
```

It is obvious that all we will get from the division is a succession of threes. This is an example of a recurring decimal, and in order to prevent endless repetition, the answer is written

$$0·\dot{3} ∴ \tfrac{1}{3} = 0·\dot{3}$$

For practical purposes we never need recurring decimals; what we generally need, is an answer given to a certain number of decimal places.

Exercise 1.16
Convert the following to decimals, correcting the answers, where necessary, to 4 decimal places.
1. $\frac{1}{4}$ 2. $\frac{3}{4}$ 3. $\frac{3}{8}$ 4. $\frac{11}{16}$
5. $\frac{1}{2}$ 6. $\frac{2}{3}$ 7. $\frac{21}{32}$ 8. $\frac{39}{64}$
9. $1\frac{5}{6}$ 10. $2\frac{7}{16}$

1.3 The metric system

The metric system is essentially a decimal system. In this chapter we shall deal only with the metric system as it applies to the measurement of length.

The standard of length is the metre (abbreviation: m) but for many engineering purposes the metre is too large a unit and hence it is split up into smaller units as follows:

$$1 \text{ metre (m)} = 10 \text{ decimetres (dm)}$$
$$= 100 \text{ centimetres (cm)}$$
$$= 1000 \text{ millimetres (mm)}$$

Most engineering drawings will be dimensioned in millimetres, but for very long lengths the metre will be used. We may, for instance, have a length of 357·632 m, meaning 357 metres and 632 millimetres.

Examples

1. Convert 7·327 m into millimetres.
   All we have to do is to multiply 7·327 by 1000. Thus,
   $$7·327 \text{ m} = 7·327 \times 1000 = \underline{7327 \text{ mm}}$$
2. Convert 9394 mm into metres.
   Here we divide 9394 by 1000, thus
   $$9394 \text{ mm} = \frac{9394}{1000} = \underline{9·394 \text{ m}}$$

For very accurate measurements, decimal parts of a millimetre are used. A typical dimension would be 39·45 mm.

Exercise 1.17

1. Convert a measurement of 7·316 m into
   (a) millimetres   (b) centimetres
2. Convert a measurement of 5216 mm into
   (a) centimetres   (b) metres
3. Convert a length of 80·4 cm into millimetres.
4. Convert a length of 958·2 mm into centimetres.
5. How many millimetres are there in 8·48 metres?

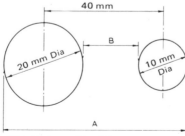

Fig. 1.2

PRACTICAL EXAMPLES OF MEASUREMENT IN THE METRIC SYSTEM

1. Find the dimensions A and B in Fig. 1.2.
   $$A = 40 + \frac{20}{2} + \frac{10}{2} = 40 + 10 + 5 = \underline{55 \text{ mm}}$$
   $$B = 40 - \frac{20}{2} - \frac{10}{2} = 40 - 10 - 5 = \underline{25 \text{ mm}}$$
2. Find the dimension A shown in Fig. 1.3.

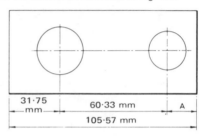

Fig. 1.3

It is standard practice on the Continent to use a comma as the decimal marker. In the United Kingdom, for the present, the accepted practice is to use the full point, preferably placed centrally (as in this book); for example, 3·2, but also sometimes on the base line, as 3.2, where this is more convenient.

3. A steel bar is 1803·4 mm long. How many lengths 89 mm long can be cut from it and what length remains?

   $$\text{Number of lengths} = \frac{1803·4}{89} \doteqdot 20 \text{ whole lengths}$$

   $$\text{Length remaining} = 1803·4 - 20 \times 89$$
   $$= 1803·4 - 1780 = \underline{23·4 \text{ mm}}$$

4. 6 turns of a screw drive it 3 mm.
   (a) How far does 25 turns drive it?
   (b) How many turns are needed to drive it 20 mm?

   (a) 6 turns drive the screw 3 mm
        1 turn drives the screw $3 \div 6 = 0·5$ mm
        25 turns drive the screw $25 \times 0·5 = \underline{12·5 \text{ mm}}$

   (b) 1 turn drives the screw 0·5 mm
        Number of turns needed to drive the screw 20 mm
        $= 20 \div 0·5 = \underline{40}$

Exercise 1.18

1. Calculate the dimension L in Fig. 1.4.
2. Calculate the dimension X in Fig. 1.5.
3. Find the dimension marked A on the shaft shown in Fig. 1.6(a).

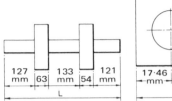

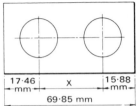

Fig. 1.4   Fig. 1.5

4. Find the dimensions A, B and C in Fig. 1.6(b). (The diagram shows two rollers resting on a surface table.)
5. Fig. 1.7(a) shows a method used for measuring the diameter of a hole. Find dimensions A and B.
6. Fig. 1.7(b) shows a bar resting in a vee block. Calculate the dimension H.
7. Calculate the dimensions A and B for the countersunk head shown in Fig. 1.8(a).

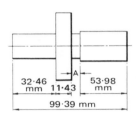

(a)

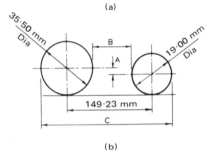

(b)

All dimensions in millimetres
Fig. 1.6

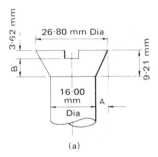

(a)

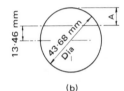

(b)

All dimensions in millimetres
Fig. 1.8

8. Fig. 1.8(b) shows a cylindrical component with a flat milled on it. Find dimension A.
9. Find dimension L for the component shown in Fig. 1.9.

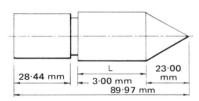

Fig. 1.9

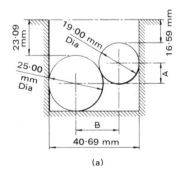

(a)

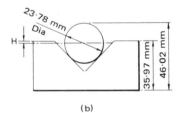

(b)

All dimensions in millimetres
Fig. 1.7

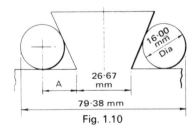

Fig. 1.10

10. Fig. 1.10 shows a method used for checking a dovetail by means of rollers. Calculate the dimension A.

11. Riverts are placed 35 mm apart in a plate 560 mm long. The distance between the centres of the first and last rivets and the edge of the plate is 16 mm. Calculate the number of rivets required.

12. A job has to be turned from a bar of steel 25·00 mm diameter. If the finished size of the bar is to be 19·53 mm, find how many roughing cuts 1·25 mm deep must be taken. What will be the depth of the finishing cut?

13. 8 turns of a screw drive it 3·2 mm. Find
    (a) How far 27 turns drive it.

(b) How many turns are needed to drive it 48·8 mm.

14. Find the distance between the insides of two holes bored at 92 mm centres and having diameters of 31·7 mm and 52·4 mm respectively.

15. A milling cutter is set to advance 0·9 mm per revolution. Find how far the cutter advances in 120 revolutions.

16. A drilling machine has a feed of 0·5 mm per revolution. Find the number of revolutions required to drill a hole 38 mm deep.

17. 15 equally spaced holes are required in a plate 558 mm long. The distance between the centres of the first and last holes and the edge of the plate has to be 20 mm. Find the spacing of the holes.

1.4   Averages, ratio, proportion and percentages

Averages. To find the average of a set of quantities, add the quantities together and divide by the number of quantities in the set.

If a boy makes the following scores at cricket: 8, 20, 3, 0, 5, 9, 15 and 12, in all he has made 72 runs in 8 innings. His *average* is said to be 72÷8 = 9. This means that if he had made 9 runs in each of his innings, his total for the 8 innings would have been 72.

You will notice that we have added all the runs together and divided by the number of innings, in order to find the average. In this way, we can find the average of any set of quantities.

Example. Find the average diameter of five components which have been ground to the following dimensions: 44·83 mm, 44·82 mm, 45·02 mm, 44·77 mm, 44·56 mm To find the average of these diameters, add the five numbers together and divide by 5.

$$\begin{array}{r} 44·83 \\ 44·82 \\ 45·02 \\ 44·77 \\ 44·56 \\ \hline 5)\overline{224·00} \\ \hline 44·80 \end{array}$$

The average diameter of the components is 44·80 mm.

Example. Find the average size of 15 diameters if 5 have a dimension of 24·99 mm, 6 have a dimension of 25·01 mm and the remaining 4 have a dimension of 25·00 mm.

Total size of 5 diameters at
$$24·99 \text{ mm} = 5 \times 24·99 = 124·95$$
Total size of 6 diameters at
$$25·01 \text{ mm} = 6 \times 25·01 = 150·06$$
Total size of 4 diameters at
$$25·00 \text{ mm} = 4 \times 25·00 = \underline{100·00}$$
Total size of 15 diameters $= 375·01$

Average size $= \dfrac{375·01}{15} = \underline{25·00 \text{ mm}}$ (correct to 2 places of decimals).

Exercise 1.19

1. Find the average of the following numbers: 95, 128, 38, 97 and 217.

2. Find the average of the following readings: 56·64 mm, 57·15 mm, 57·40 mm, 56·13 mm.

3. The temperature of a heat treatment furnace was taken at regular intervals of time. The following temperatures were recorded: 723, 731, 734, 735, 733, 736, 730 and 726 degrees Celsius. Calculate the average temperature during the period.

4. The speed of a shaping machine ram is 21 m/min during the cutting stroke and 33 m/min during the return stroke. Calculate the average speed of the ram.

5. In measuring the diameter of a rod the following readings were taken at different points along its length: 33·53, 33·43, 33·48 and 33·18 mm. Find the average diameter of the rod.

6. Find the average mass of 22 castings if 11 castings have a mass of 12 kg each and 8 have a mass of $12\frac{1}{2}$ kg each.

7. 15 components have an average mass of 3·2 kg. If 5 more components having an average mass of 3·05 kg are added to the previous 15, what is now the average mass of the components?

8. In a certain machine shop, soluble oil is used as a coolant. The amounts used in six successive weeks are as follows: 885 litres, 730 litres, 653 litres, 303 litres, 235 litres and 510 litres. Calculate the average amount of coolant used per week.

9. In a production machine shop, the numbers of components produced were as follows: 326 in 5 days; 408 in 6 days; 205 in 3 days; 128 in 2 days. Find the average number of components produced per day.

10. For the purpose of calculating the time taken to produce 1000 components, three men are chosen and timed whilst producing one component. The times recorded were as follows: 5·4 hours, 5·8 hours and 5·3 hours.
    (a) Estimate the total time required to produce the 1000 components.
    (b) If 8 men are to work on producing the components, estimate the time needed to produce all the components.

Ratio. In Fig. 1.11 a line AB which is 75 mm long has been drawn. The line has been divided at C, so that AC is 25 mm long. We can say at once that AC is $\frac{1}{2}$ of BC. We can put this in another way, by stating that the line AB has been divided into two parts in the ratio of 1 is to 2. This is commonly written as 1:2, the two dots, one under the other, replacing the words 'is to'. The two ways of expressing this, $\frac{1}{2}$ and 1:2 mean exactly the same. We know, from our work with fractions, that $\frac{1}{2}$ means 1÷2. Therefore the sign : means division.

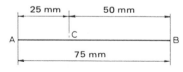

Fig. 1.11

The ratio 20:5 means the number of times 5 is contained in 20. It is the fraction $\frac{20}{5}$ which is equal to 4. Notice that the first number stated is the numerator and the second

number is the denominator of the fraction. The fraction $\frac{3}{4}$ means 3:4 and the ratio 6:5 means $\frac{6}{5}$.

Exercise 1.20

Express the following fractions as ratios:
1. $\frac{2}{3}$ 2. $\frac{5}{8}$ 3. $\frac{9}{7}$ 4. $\frac{5}{2}$

Express the following ratios as fractions in their lowest form:
5. 4:5   6. 8:3   7. 2:9   8. 4:6
9. 12:4   10. 10:3   11. 8:12   12. 9:15

You will notice from Fig. 1.11 that AC = $\frac{1}{3}$ of AB, whilst BC = $\frac{2}{3}$ of AB. That is, if we divide AB into 3 equal parts then AC is equal to 1 of those parts and BC is equal to 2 of the parts. You will remember that AB was divided into 2 parts in the ratio of 1:2. It is clear therefore that the ratio 1:2 means that the line has been divided into 1+2 = 3 parts, and that one part is $\frac{1}{3}$ and the other $\frac{2}{3}$. The following examples show the method you should use.

Examples

1. Divide a line 350 mm long into two parts in the ratio 4:3.

    Total number of parts   = 4+3   = 7
    Length of each part   = 350÷7 = 50 mm
    Length of one part   = 4×50 = 200 mm
    Length of the other part = 3×50 = 150 mm

2. Divide a line 480 mm long into three parts in the ratio 3:4:5.

    Total number of parts   = 3+4+5 = 12
    Length of each part   = 480÷12 = 40 mm
    Length of first part   = 3×40 = 120 mm
    Length of second part   = 4×40 = 160 mm
    Length of third part   = 5×40 = 200 mm

Sometimes ratios are stated in either fractional or decimal form. In these cases we can use similar methods to those shown in the previous two examples.

Example. An alloy consists of copper and zinc in the ratio 3·25:4·50 by mass. Calculate the masses of copper and zinc in a sample of the alloy which has a mass of 1·55 kg.

    Total number of parts = 3·25+4·50 = 7·75
    Mass of each part   = 1·55÷7·75 = 0·2 kg
    Mass of copper   = 3·25×0·2 = 0·65 kg
    Mass of zinc   = 4·50×0·2 = 0·90 kg

Example. A bar of steel 900 mm long is to be cut into three parts in the ratio 0·5:1·75:3·75. Find the length of each part.

    Total number of parts = 0·5+1·75+3·75 = 6·0
    Length of each part   = 900÷6   = 150 mm
    Length of first part   = 0·5×150 = 75 mm
    Length of second part = 1·75×150 = 262·5 mm
    Length of third part   = 3·75×150 = 562·5 mm

Exercise 1.21

1. Divide a line 1000 mm long into two parts in the ratio 3:2.
2. Divide a line 160 mm long into three parts in the ratio 1:2:5.
3. A brass consists of 3 parts copper to 2 parts zinc. Find the masses of copper and zinc in a casting having a mass of 20 kg.
4. A right-angled triangle has sides in the ratio 3:4:5. Calculate the length of each side if the total length of all the sides is 360 mm.
5. A bronze consists of copper, lead and tin in the ratios of 7:2:1 by mass. Find the mass of each metal in a bronze bearing which has a mass of 14 kg.
6. A white metal bearing consists of 3·25 parts of copper to 2·05 parts of tin. Calculate the masses of copper and tin in a bearing which has a mass of 477 grams.
7. Divide a line 210 mm long into 3 parts in the ratio 1·5:2·5:3.
8. The lengths of two of the sides of a triangle are in the ratio $1\frac{1}{2}:2\frac{1}{4}$. If the length of the two sides added together is 75 mm, find the length of each side.
9. The production from three machines producing identical components is in the ratio 1:1·15:1·25. If the total number of components produced in a week is 680, find the number produced on each machine.
10. A bar of Babbitt metal consists of 2 parts antimony, 4 parts copper and 19 parts tin. Find the mass of each metal if the total mass of the bar is 1·25 kg.

Proportion. Proportion is an extension of our work with ratios. If 25 litres of oil have a mass of 20 kg then it is clear that 50 litres of oil will have a mass of 40 kg. That is, if we double the quantity of oil then we double the mass. In the same way, $12\frac{1}{2}$ litres of oil have a mass of 10 kg. That is, half the quantity of oil has a mass only half as much. This is an example of *direct proportion*.

We can always use ratios in problems on proportion as the following example shows.

Example. If a casting with a mass of 6000 kg costs £420, what would be the cost of a casting having a mass of 10 000 kg, at the same rate?
The mass is increased in the ratio $\frac{10\,000}{6000} = \frac{10}{6}$
∴ Cost of a casting having a mass of 10 000 kg
$$= 420 \times \tfrac{10}{6} = £700.$$
In the workshop, proportion can be used when dealing with tapers. Tapers are often expressed as a ratio (e.g. 1 in 20).

Example. A tapered piece has a taper of 1 in 20 on diameter and it has a length of 240 mm. If the smaller diameter is 28 mm, find the larger diameter.
A taper of 1 in 20 means a taper of 1 mm in a length of 20 mm. Therefore, in a length of 240 mm there will be a taper of $\frac{240}{20} = 12$ mm. Hence the larger diameter is 28+12 = 40 mm.

Example. Fig. 1.12 shows a taper piece. Find the taper ratio.

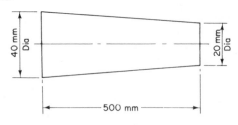

Fig. 1.12

Difference in diameters = 40−20 = 20 mm
Length = 500 mm
$$\text{Taper ratio} = \frac{\text{difference in diameters}}{\text{length of taper}}$$
$$= \tfrac{20}{500} = \tfrac{1}{25}$$
∴ Taper ratio is 1 in 25.

So far we have dealt only with direct proportion. Suppose 8 men working on a certain job take 10 days to complete it. If we double the number of men then we halve the time taken. If we halve the number of men then we double the time. This is called *inverse proportion*.

Example. 20 men operating capstans produce 3000 components in 12 working days. Find how long it would take 15 men to produce 3000 components.
The number of men is reduced in the ratio $\frac{15}{20}$
The number of days is increased in the ratio $\frac{20}{15}$
∴ Number of days required = $\frac{20}{15} \times 12 = 16$ days.

Example. Two pulleys 150 mm and 50 mm diameter respectively are connected by a belt. If the larger pulley revolves at 80 rev/min find the speed of the smaller pulley.
The smaller pulley must revolve faster than the larger pulley, so this is another example of inverse proportion.
The pulley diameters are reduced in the ratio $\frac{50}{150} = \frac{1}{3}$
The speed is increased in the ratio $\frac{3}{1} = 3$
∴ Speed of smaller pulley = 3×80 = 240 rev/min.

Exercise 1.22

1. The carriage of metal billets costs £2 for 3500 kg. How much will it cost to transport 70 000 kg at the same rate?
2. The cost of a casting having a mass of 200 kg is £6. Find the cost of a casting having a mass of 50 kg at the same rate.
3. A bar of metal 3·7 m long has a mass of 111 kg. Find the mass of a similar bar 1·6 m long.

4. The feed of a lathe tool is 40 mm for 20 revolutions of the spindle. Find the feed for 37 revolutions. Find also, the number of revolutions required for a feed of 90 mm.
5. A pin has a taper of 1 in 30 on diameter. It has a length of 90 mm. If the smaller diameter is 12 mm, find the larger diameter.
6. The following dimensions refer to a taper piece: length 500 mm; small diameter 20 mm; large diameter 30 mm. Find the taper ratio.
7. The taper on the spindle nose of a machine is 7 in 24 on diameter. The large diameter is 70 mm and the small diameter is 32 mm. Find the length of the taper.
8. 10 men produce 500 components in 5 working days. How long would it take 20 men to produce 2000 components?
9. Two pulleys 300 mm diameter and 50 mm diameter respectively are connected by a belt. If the smaller pulley revolves at 240 rev/min, find the speed of the larger pulley.
10. Two gear wheels mesh together. One has 40 teeth and revolves at 100 rev/min. The other has 25 teeth. Find its speed.

Percentages. From your work with fractions and decimals, you will realize that it is possible to express any fraction with denominators 10, 100, 1000 and so on. When comparing ratios and fractions it is often convenient to express them as fractions with denominators of 100. Thus,

$$\frac{1}{2} = \frac{50}{100}; \quad \frac{2}{5} = \frac{40}{100}; \quad \frac{3}{4} = \frac{75}{100}$$

It is now easy to compare $\frac{1}{2}$ with $\frac{2}{5}$ and $\frac{3}{4}$.

Fractions expressed with a denominator of 100 are called *percentages*. Thus, $\frac{1}{2} = 50$ per cent, $\frac{2}{5} = 40$ per cent and $\frac{3}{4} = 75$ per cent. The sign % is often used instead of the words per cent.

Example. Express the following fractions as percentages: (a) $\frac{9}{20}$ (b) $\frac{21}{25}$ (c) $\frac{1}{4}$.

(a) $\frac{9}{20} = \frac{45}{100} = \underline{45\%}$
(b) $\frac{21}{25} = \frac{84}{100} = \underline{84\%}$
(c) $\frac{1}{4} = \frac{25}{100} = \underline{25\%}$

To express a percentage as a fraction we need only divide the percentage by 100.

Example. Express the following percentages as fractions: (a) 45% (b) 70% (c) 52%.

(a) $45\% = \frac{45}{100} = \frac{9}{20}$ (cancelling by 5)
(b) $70\% = \frac{70}{100} = \frac{7}{10}$ (cancelling by 10)
(c) $52\% = \frac{52}{100} = \frac{13}{25}$ (cancelling by 4)

So far, we have dealt only with the conversion of fractions into percentages and percentages into fractions. It is often necessary to convert decimals into percentages and percentages into decimals. Thus,

$$0\cdot3 = \frac{3}{10} = \frac{30}{100} = 30\%$$
$$0\cdot28 = \frac{28}{100} = 28\%$$

It is clear, therefore, that to convert a decimal to a percentage all we have to do is to multiply the decimal by 100. That is, we shift the decimal point two places to the right. Thus,

$$0\cdot41 = (0\cdot41\times100)\% = 41\%$$
$$0\cdot268 = (0\cdot268\times100)\% = 26\cdot8\%$$
$$0\cdot5372 = (0\cdot5372\times100)\% = 53\cdot72\%$$

Obviously then, to convert a percentage to a decimal we divide by 100. Thus,

$$71\% = \frac{71}{100} = 0\cdot71$$
$$26\% = \frac{26}{100} = 0\cdot26$$

Dividing by 100 is, of course, the same as shifting the decimal point two places to the left. Thus,

$$48\cdot7\% = \frac{48\cdot7}{100} = 0\cdot487$$

$$91\cdot83\% = \frac{91\cdot83}{100} = 0\cdot9183$$

Exercise 1.23
Convert the following decimals to percentages:
1. 0·5   2. 0·63   3. 0·813   4. 0·667   5. 0·7235
Convert the following percentages to decimals:
6. 40%   7. 65%   8. 48·2%   9. 6%   10. 0·3%

PRACTICAL EXAMPLES OF PERCENTAGES
1. What is 15% of 80?
   15% of 80 = $\frac{15}{100}$ of 80 = $\frac{15}{100}\times80 = \underline{12}$
2. What percentage of 120 is 20?
   $\frac{20}{120} = \frac{1}{6} = 0\cdot167 = (0\cdot167\times100)\% = \underline{16\cdot7\%}$
3. The output of a machine producing 550 components per day is to be increased by 8%. Calculate the new output.
   100% = 550 components
   1% = $\frac{550}{100}$ = 5·5 components
   8% = 8×5·5 = 44 components
   ∴ New output = 550+44 = <u>594 components.</u>
4. In 40 kg of brass there are 10 kg of zinc, and the remainder is copper. Find the percentages of zinc and copper in the brass.
   In 40 kg of brass there are 10 kg of zinc.
   In 1 kg of brass there are $\frac{10}{40}$ kg of zinc.
   In 100 kg of brass there are $\frac{10}{40}\times100$ = 25 kg of zinc.
   ∴ Percentage of zinc = <u>25%.</u>
   Percentage of copper = 100−25 = <u>75%</u>
5. What is 7% of 112 kg?
   7% of 112 = $\frac{7}{100}\times112 = \frac{784}{100} = \underline{7\cdot84\ kg}$
6. In an alloy there are 5 parts copper and 20 parts tin. Calculate the percentage composition of the alloy.
   The total number of parts is 5+20 = 25
   In 25 parts there are 5 parts copper.
   In 1 part there are $\frac{5}{25}$ parts copper.
   In 100 parts there are $\frac{5}{25}\times100$ = 20 parts copper.
   ∴ Percentage of copper = <u>20%</u>
   Percentage of tin = 100−20 = <u>80%</u>

7. A bronze bearing consists of 70% copper, 20% lead and the remainder tin. Find the mass of each metal if the bearing has a mass of 200 kg.

70% of 200 = $\frac{70}{100}$ of 200 = $\frac{70}{100} \times 200$ = 140 kg

20% of 200 = $\frac{20}{100}$ of 200 = $\frac{20}{100} \times 200$ = 40 kg

∴ Mass of copper = 140 kg

Mass of lead = 40 kg

Mass of tin = 200−140−40 = 200−180 = 20 kg

Exercise 1.24

1. (a) What is 12% of 75?
    (b) What is 20·5% of 105?

2. (a) What percentage of 150 is 24?
    (b) What percentage of 180 is 30?
    (c) What percentage of 40 is 15?

3. On a capstan lathe, 3% of the components produced are scrapped because of defective workmanship. If 900 components are produced in a week, how many are scrapped?

4. The composition of an alloy is 44 parts copper, 14 parts tin and 2 parts antimony. What is the percentage of each metal?

5. A man has to make 45 kg of a lead-tin alloy for experimental use. The alloy must possess 60% lead and 40% tin. Find the masses of each of the two metals that must be used.

6. How much copper must be melted with 50 kg zinc in order to make an alloy consisting of 75% copper and 25% zinc? What is the total mass of the alloy produced.

7. A factory employing 250 men wishes to increase its employees by 12%. How many extra men must be employed?

8. In a sample of iron ore 20% is iron. How much of the ore is needed to yield 8000 kg of iron?

9. A machine costing £500 depreciates 5% of its value during each year of its working life. Find its value after 2 years.

10. In a lathe, 18% of the power supplied by the motor is lost due to friction. If the motor supplies 5 kW calculate the loss of power due to friction.

1.5 Squares, square roots and reciprocals

Squares of numbers. When a number is multiplied by itself the result is called the *square* of the number. Thus the square of $12 = 12 \times 12 = 144$. Instead of writing 12×12 we often write 12^2 which is read as the square of 12. Thus,

$$13^2 = 13 \times 13 = 169$$
$$(1 \cdot 28)^2 = 1 \cdot 28 \times 1 \cdot 28 = 1 \cdot 638$$

The square of any number can be found by multiplication but a great deal of effort is saved by using printed tables. Part of the tables of squares is shown below. In this table the squares of numbers are given *correct to 4 significant figures*.

HOW TO USE THE TABLES OF SQUARES OF NUMBERS

1. *The square of a number having two significant figures.* To find $(1 \cdot 8)^2$, find $1 \cdot 8$ in the first column (see sample table) and move along this row to the number under the column headed 0. We find this number to be $3 \cdot 240$.
   Thus $(1 \cdot 8)^2 = 3 \cdot 240$
   Similarly, $(2 \cdot 2)^2 = 4 \cdot 840$

2. *The square of a number having three significant figures.* To find $(2 \cdot 15)^2$ find $2 \cdot 1$ in the first column (see sample table) and move along this row to the number under the column headed 5. We find this number to be $4 \cdot 623$.
   Thus $(2 \cdot 15)^2 = 4 \cdot 623$
   Similarly, $(1 \cdot 98)^2 = 3 \cdot 920$

3. *The square of a number having four significant figures.* To find $(2 \cdot 018)^2$ find $2 \cdot 0$ in the first column (see sample table) and move along this row to the number under the column headed 1. We find this number to be $4 \cdot 040$. Now move along the same row to the number under the column headed 8 of *proportional parts* and read 33. Add this 33 (that is $0 \cdot 033$) to $4 \cdot 040$ to give $4 \cdot 073$.
   Thus $(2 \cdot 018)^2 = 4 \cdot 073$
   Similarly, $(1 \cdot 356)^2 = 1 \cdot 839$

4. *Squares of numbers outside the range of the tables.* Although the tables only give the squares of numbers from 1 to 10 they can be used to find the squares of numbers outside this range. The method is shown in the following examples.

SQUARES

	0	1	2	3	4	5	6	7	8	9	Proportional Parts								
											1	2	3	4	5	6	7	8	9
1·0	1·000	1·020	1·040	1·061	1·082	1·103	1·124	1·145	1·166	1·188	2	4	6	8	10	13	15	17	19
1·1	1·210	1·232	1·254	1·277	1·300	1·323	1·346	1·369	1·392	1·416	2	5	7	9	11	14	16	18	21
1·2	1·440	1·464	1·488	1·513	1·538	1·563	1·588	1·613	1·638	1·664	2	5	7	10	12	15	17	20	22
1·3	1·690	1·716	1·742	1·769	1·796	1·823	1·850	1·877	1·904	1·932	3	5	8	11	13	16	19	22	24
1·4	1·960	1·988	2·016	2·045	2·074	2·103	2·132	2·161	2·190	2·220	3	6	9	12	14	17	20	23	26
1·5	2·250	2·280	2·310	2·341	2·372	2·403	2·434	2·465	2·496	2·528	3	6	9	12	15	19	22	25	28
1·6	2·560	2·592	2·624	2·657	2·690	2·723	2·756	2·789	2·822	2·856	3	7	10	13	16	20	23	26	30
1·7	2·890	2·924	2·958	2·993	3·028	3·063	3·098	3·133	3·168	3·204	3	7	10	14	17	21	24	28	31
1·8	3·240	3·276	3·312	3·349	3·386	3·423	3·460	3·497	3·534	3·572	4	7	11	15	18	22	26	30	33
1·9	3·610	3·648	3·686	3·725	3·764	3·803	3·842	3·881	3·920	3·960	4	8	12	16	19	23	27	31	35
2·0	4·000	4·040	4·080	4·121	4·162	4·203	4·244	4·285	4·326	4·368	4	8	12	16	20	25	29	33	37
2·1	4·410	4·452	4·494	4·537	4·580	4·623	4·666	4·709	4·752	4·796	4	9	13	17	21	26	30	34	39
2·2	4·840	4·884	4·928	4·973	5·018	5·063	5·108	5·153	5·198	5·244	4	9	13	18	22	27	31	36	40
2·3	5·290	5·336	5·382	5·429	5·476	5·523	5·570	5·617	5·664	5·712	5	9	14	19	23	28	33	38	42
2·4	5·760	5·808	5·856	5·905	5·954	6·003	6·052	6·101	6·150	6·200	5	10	15	20	24	29	34	39	44

Examples

1. To find $(395 \cdot 6)^2$

   $(395 \cdot 6)^2 = (3 \cdot 956 \times 100)^2 = 3 \cdot 956 \times 100 \times 3 \cdot 956 \times 100$
   $= (3 \cdot 956)^2 \times 100^2$

   Now look up $(3 \cdot 956)^2$ in the tables. It will be found to be $15 \cdot 65$

   $\therefore (395 \cdot 6)^2 = 15 \cdot 65 \times 100^2 = \underline{156\ 500}$

2. To find $(0 \cdot 1529)^2$

   Now $(0 \cdot 1529)^2 = \left(\dfrac{1 \cdot 529}{10}\right)^2 = \dfrac{1 \cdot 529}{10} \times \dfrac{1 \cdot 529}{10}$

   $= \dfrac{(1 \cdot 529)^2}{100}$

   Now look up $(1 \cdot 529)^2$ in the tables. It will be found to be $2 \cdot 338$

   $\therefore (0 \cdot 1529)^2 = \dfrac{(1 \cdot 529)^2}{100} = \dfrac{2 \cdot 338}{100} = \underline{0 \cdot 023\ 38}$

When the tables are to be used to find the square of a number with more than 4 significant figures, correct the number to 4 significant figures before using the tables.

Example. To find $(2 \cdot 1568)^2$

The number correct to 4 significant figures is $2 \cdot 157$. Now use the tables to find $(2 \cdot 157)^2 = 4 \cdot 653$

   $\therefore (2 \cdot 1568)^2 = \underline{4 \cdot 653}$ correct to 4 significant figures.

Square roots. The square root of a number is the number whose square is the given number. Thus,

$$5^2 = 25 \quad \sqrt{25} = 5$$
$$6^2 = 36 \quad \sqrt{36} = 6$$

The square root of any number can be calculated by arithmetic but the method is tedious and takes quite a long time. In most cases the square root of any number may be found to sufficient accuracy by using the printed tables of square roots.

There are two tables of square roots. One gives the square roots of numbers from 1 to 10 and the other gives the square roots of numbers from 10 to 100. The reason for having two tables is as follows:

$$\sqrt{2 \cdot 5} = 1 \cdot 581$$
$$\sqrt{25} = 5$$

Thus there are two square roots for the same figures depending on the position of the decimal point. The square root tables are used in the same way as the tables of squares.

Examples

1. To find $\sqrt{8 \cdot 734}$

   $$\sqrt{8 \cdot 734} = \underline{2 \cdot 955}$$

2. To find $\sqrt{15 \cdot 62}$

   $$\sqrt{15 \cdot 62} = \underline{3 \cdot 953}$$

Although the two sets of square root tables give the square roots of numbers from 1 to 10 and 10 to 100 we can also find the square roots of numbers outside this range by using these tables. We must be sure, however, that we use the correct table and position the decimal point correctly. The methods shown in the following examples should be used.

Examples

1. To find $\sqrt{951 \cdot 6}$

   Mark off the figures in pairs to the *left* of the decimal point. Each pair of figures is called a period. Thus $951 \cdot 6$ becomes $9'51 \cdot 6$. The first period is 9 so look up $\sqrt{9 \cdot 516} = 3 \cdot 085$. To position the decimal point in the answer remember that for each period to the left of the decimal point there will be one figure to the left of the decimal point in the answer

   $$\overset{3\quad 0 \cdot 85}{9'\,51 \cdot 6}$$

   $$\therefore \sqrt{951 \cdot 6} = \underline{30 \cdot 85}$$

2. To find $\sqrt{295\ 300}$

   Marking off in periods $295\ 300$ becomes $29'53'00$. By using the tables $\sqrt{29 \cdot 5300} = 5 \cdot 434$

   $$\overset{5\quad 4\quad 3 \cdot 4}{29\ \ 53\ \ 00}$$

   $$\therefore \sqrt{295\ 300} = \underline{543 \cdot 4}$$

3. To find $\sqrt{0 \cdot 000\ 845\ 2}$

   In the case of decimal numbers mark off in periods to the right of the decimal point. $0 \cdot 000\ 845\ 2$ becomes $00'08'45'20$. Apart from the zero pair the first period is 08 so look up $\sqrt{8 \cdot 452} = 2 \cdot 907$. For each zero pair in the original number there must be one zero following the decimal point in the answer.

   $$\overset{0 \cdot\ 0\quad 2\quad 9\quad 0\quad 7}{0 \cdot 00'\,08'\,45'\,20'\,00}$$

   $$\therefore \sqrt{0 \cdot 000\ 845\ 2} = \underline{0 \cdot 029\ 07}$$

4. To find $\sqrt{0 \cdot 000\ 062\ 17}$

   Marking off in pairs to the right of the decimal point $0 \cdot 000\ 062\ 17$ becomes $00'00'62'17$. Apart from the zero pairs the first pair is 62 so look up $\sqrt{62 \cdot 17} = 7 \cdot 884$.

   $$\overset{0 \cdot\ 0\quad 0\quad 7\quad 8\quad 8\quad 4}{0 \cdot 00'\,00'\,62'\,17'\,00'\,00}$$

   Thus $\sqrt{0 \cdot 000\ 062\ 17} = \underline{0 \cdot 007\ 884}$

Reciprocals of numbers. The reciprocal of a number is

$$\dfrac{1}{\text{number}}$$

The reciprocal of 8 is $\dfrac{1}{8}$

The reciprocal of $22 \cdot 4$ is $\dfrac{1}{22 \cdot 4}$

The tables of reciprocals are used in a similar way to the tables of squares. The tables give the reciprocals of numbers from 1 to 10 in decimal form. From the tables:

the reciprocal of   7 $= 0·1429$
the reciprocal of $2·918 = 0·3427$

The method of finding the reciprocals of numbers less than 1 or greater than 10 is shown in the following examples.

Examples

1. To find $\dfrac{1}{563·2}$

$$\frac{1}{563·2} = \frac{1}{5·632} \times \frac{1}{100} = \frac{0·1775}{100} = \underline{0·001\,775}$$

2. To find $\dfrac{1}{0·2968}$

$$\frac{1}{0·2968} = \frac{1}{2·968} \times 10 = 0·3369 \times 10 = \underline{3·369}$$

Use of tables in calculations. Calculations may often be speeded up by making use of the tables of squares, square roots and reciprocals.

Examples

1. To find the value of $\sqrt{(8·135)^2 + (12·36)^2}$

$$\sqrt{(8·135)^2 + (12·36)^2} = \sqrt{66·18 + 152·8}$$
$$\text{(by using table of squares)}$$
$$= \sqrt{218·98}$$
$$= \underline{14·80}$$
$$\text{(by using square root tables)}$$

2. To find the value of $\dfrac{1}{3·52} + \dfrac{1}{10·83} + \dfrac{1}{17·62}$

$$\frac{1}{3·52} + \frac{1}{10·83} + \frac{1}{17·62} = 0·2841 + 0·0923 + 0·0568$$
$$= \underline{0·4332}$$

Exercise 1.25

1. Find the squares of the following numbers:

(a) 2·5	(f) 38	(k) 0·025
(b) 4·16	(g) 49·7	(l) 0·263
(c) 3·623	(h) 160·2	(m) 0·003 817
(d) 7·782	(i) 2184	(n) 0·3918
(e) 5·168	(j) 16·23	(o) 0·069 75

2. Find the square roots of the numbers given in question 1.

3. Find the reciprocals of the numbers given in question 1.

4. Find values of the following using the tables of squares, square roots and reciprocals.

(a) $\sqrt{(2·65)^2 + (5·16)^2}$

(b) $\sqrt{(17·42)^2 + (23·61)^2}$

(c) $\sqrt{(11·18)^2 - (5·23)^2}$

(d) $\sqrt{(30·15)^2 - (18·17)^2}$

(e) $\dfrac{1}{3·5} + \dfrac{1}{7·7}$

(f) $\dfrac{1}{70·16} + \dfrac{1}{18·24}$

(g) $\dfrac{1}{0·081\,73} + \dfrac{1}{0·2618}$

(h) $\dfrac{1}{17·65} - \dfrac{1}{80·55}$

(i) $\dfrac{1}{\sqrt{7·517}} + \dfrac{1}{(3·625)^2}$

(j) $\dfrac{1}{7·136} + \dfrac{1}{\sqrt{80·32}} + \dfrac{1}{(2·561)^2}$

1.6 Directed numbers

Positive and negative numbers. Fig. 1.13 shows part of a Celsius thermometer. The freezing point of water is 0 °C. Temperatures higher than freezing may be read off the scale directly and so may those below freezing. We now need a way of indicating whether a temperature is

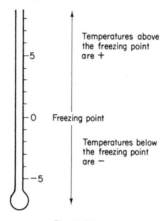

Fig. 1.13

above or below freezing. We might say that a temperature is 8 °C above freezing or a temperature is 3 °C below freezing. These statements however are not compact enough for calculations. We therefore say that a temperature of +8 °C is a temperature of 8 °C *above* freezing point. A temperature of −3 °C means that the temperature is 3 °C *below* freezing point. Thus we have used the signs + and − to indicate a *change of direction*.

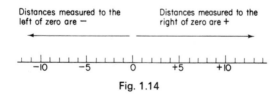

Distances measured to the left of zero are −

Distances measured to the right of zero are +

Fig. 1.14

Starting from a given point (zero in Fig. 1.14), distances measured to the right may be regarded as being positive distances whilst distances measured to the left may be regarded as negative distances.

Numbers which have a sign attached to them are called *directed numbers*. Thus +6 is a positive number whilst −6 is a negative number.

Rules for the use of directed numbers

THE ADDITION OF POSITIVE AND NEGATIVE NUMBERS

1. *To find the value of +4+3.* Measure 4 units to the right of 0 and then a further 3 units to the right. The final position is 7 units to the right of 0 (Fig. 1.15).
$$\therefore \ +4+3 = \underline{+7}$$
2. *To find the value of −6−2.* Measure 6 units to the left

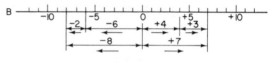

Fig. 1.15

of 0 and then a further 2 units to the left. The final position is 8 units to the left of 0 (Fig. 1.15).
$$\therefore \ -6-2 = \underline{-8}$$

From these results we obtain the rule:
To add several numbers together whose signs are the same add the numbers together. The sign of the result is the same as the sign of each of the numbers.

Positive signs are often omitted as shown in the following examples:
1. $+7+6 = +13$
   More often this is written:
   $$7+6 = 13$$
2. $-5+(-7) = -12$
   More often this is written:
   $$-5-7 = -12$$
3. $-2-3-5 = -10$

THE ADDITION OF NUMBERS HAVING DIFFERENT SIGNS

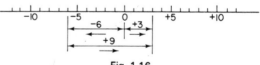

Fig. 1.16

1. *To find the value of −6+9.* Measure 6 units to the left of 0 to represent −6. From this point measure 9 units to the right to represent +9. The final result is 3 units to the right of 0 (Fig. 1.16).
$$\therefore \ -6+9 = \underline{+3}$$

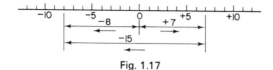

Fig. 1.17

MULTIPLICATION OF DIRECTED NUMBERS

$$4+4+4 = 12$$
that is
$$3 \times 4 = 12$$
Thus two positive numbers multiplied together give a positive result.

$$-4-4-4 = -12$$
that is
$$3 \times (-4) = -12$$
Thus a positive number multiplied by a negative number gives a negative result.

Suppose we wish to find the value of $(-3) \times (-4)$. We can write (-3) as $-(+3)$ and hence,
$$(-3) \times (-4) = -(+3) \times (-4) = -(-12) = +12$$
Thus a negative number multiplied by a negative number yields a positive result.

We may summarize the results as follows:
$$(+) \times (+) = + \qquad (-) \times (+) = -$$
$$(+) \times (-) = - \qquad (-) \times (-) = +$$
The rule is:
The product of two numbers with like signs is positive whilst the product of two numbers with unlike signs is negative.

2. To find the value of +7−15.

Measure 7 units to the right of 0 to represent +7. From this point measure 15 units to left to represent −15. The final position is 8 units to the left of 0 (Fig. 1.17). Thus,
$$+7-15 = \underline{-8}$$

From these results we obtain the rule:
To add two numbers together whose signs are different subtract the numerically smaller from the numerically larger. The sign of the result will be the same as that of the numerically larger number.

Examples
1. $-11+ 6 = \underline{-5}$
2. $12-16 = \underline{-4}$
3. $15- 9 = \underline{6}$
4. $-8+15 = \underline{7}$

When dealing with several numbers having mixed signs add the positive and negative numbers together separately. The set of numbers is then reduced to two numbers, one positive and the other negative. These are then added in the way previously shown. Thus,
$$-8+9-6+5-2 = -16+14 = \underline{-2}$$

Examples
1. $\qquad 5 \times 6 = \underline{30}$
2. $\qquad 5 \times (-6) = \underline{-30}$
3. $\qquad (-5) \times 6 = \underline{-30}$
4. $(-5) \times (-6) = \underline{30}$

DIVISION OF DIRECTED NUMBERS

The rule must be similar to that used in multiplication and it is as follows:
When dividing, numbers with like signs give a positive answer whilst numbers with unlike signs give a negative answer.
Thus: $\qquad (+) \div (+) = (+) \qquad (-) \div (+) = (-)$
$\qquad (+) \div (-) = (-) \qquad (-) \div (-) = (+)$

Examples
1. $\dfrac{15}{3} = \underline{5}$
2. $\dfrac{15}{(-3)} = \underline{-5}$
3. $\dfrac{(-15)}{3} = \underline{-5}$
4. $\dfrac{(-15)}{(-3)} = \underline{5}$

THE SUBTRACTION OF DIRECTED NUMBERS

1. *To find the value of −5−(+7).* From Fig. 1.18 it will be seen that $-(+7)$ is the same as -7 and hence,
$$-5-(+7) = -5-7 = \underline{-12}$$
2. *To find the value of +3−(−8).* From Fig. 1.18 it will be seen that $-(-8)$ is the same as $+8$ and hence,
$$+3-(-8) = 3+8 = \underline{11}$$

From these results we obtain the rule:
To subtract a directed number change its sign and add the resulting number.

Examples
1. $-8-(-6) = -8+6 = \underline{-2}$
2. $10-(-3) = 10+3 = \underline{13}$
3. $8-(+5) = 8-5 = \underline{3}$
4. $-5-(+7) = -5-7 = \underline{-12}$

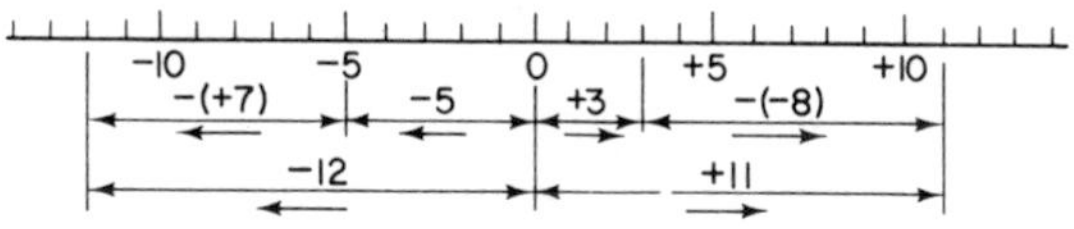

Fig. 1.18

Exercise 1.26
Find the values of the following:
1. $8+7$
2. $-13-12$
3. $-8-9$
4. $5+8$
5. $7-11$
6. $8-16$
7. $-5-12$
8. $-4+8$
9. $11-5$
10. $-8-10$
11. $-7-5-4$
12. $-6+8-3$
13. $17-8-5$
14. $20-19-8+3$

15. $8-(+5)$
16. $-4-(-7)$
17. $8-(-3)$
18. $-6-(-2)$
19. $-5-(+6)$
20. $-3-(-6)-(-4)$

21. 5×4
22. $(-5)\times4$
23. $5\times(-4)$
24. $(-5)\times(-4)$
25. $(-3)^2$
26. $(-8)^2$

27. $3\times(-4)+2\times(-3)$
28. $(-3)\times(2)-(-2)\times4$
29. $6\div3$
30. $6\div(-3)$
31. $(-6)\div(3)$
32. $(-6)\div(-3)$

33. $(-10)\div(-5)$
34. $1\div(-1)$
35. $(-1)\div1$
36. $\dfrac{(-3)\times(-4)}{(-2)}$

1.7   Use of logarithms

Workshop calculations frequently require the multiplication and division of awkward numbers. The calculation is often laborious if ordinary multiplication and division is used and the process may be greatly speeded up by using logarithms.

Reading the log tables. The log tables give the logarithms of numbers between 1 and 10. Part of a log table is shown below. The figures in the first column of the complete table are the numbers from 1 to 9·9. The corresponding figures in the column headed 0 are the logarithms of these numbers. Thus,

$$\log 5\cdot5 = 0\cdot7404$$

If the number has a *third* significant figure the logarithm is found in the appropriate column of the next nine columns. Thus,

$$\log 5\cdot57 = 0\cdot7459$$

When the number has a *fourth* significant figure we use the last nine columns which give us, for every fourth significant figure, a number which must be *added* to the logarithm already found for the first three significant figures. Thus, to find log 5·576 we find log 5·57 = 0·7459. Using the last nine columns, we find, in the column headed 6, the number 5. This is added to 0·7459 to give

$$\log 5\cdot576 = 0\cdot7459+0\cdot0005 = 0\cdot7464$$

LOGARITHMS

	0	1	2	3	4	5	6	7	8	9	1	2	3	4	5	6	7	8	9
55	·7404	7412	7419	7427	7435	7443	7451	7459	7466	7474	1	2	2	3	4	5	5	6	7
56	·7482	7490	7497	7505	7513	7520	7528	7536	7543	7551	1	2	2	3	4	5	5	6	7
57	·7559	7566	7574	7582	7589	7597	7604	7612	7619	7627	1	2	2	3	4	5	5	6	7
58	·7634	7642	7649	7657	7664	7672	7679	7686	7694	7701	1	1	2	3	4	4	5	6	7
59	·7709	7716	7723	7731	7738	7745	7752	7760	7767	7774	1	1	2	3	4	4	5	6	7
60	·7782	7789	7796	7803	7810	7818	7825	7832	7839	7846	1	1	2	3	4	4	5	6	6
61	·7853	7860	7868	7875	7882	7889	7896	7903	7910	7917	1	1	2	3	4	4	5	6	6
62	·7924	7931	7938	7945	7952	7959	7966	7973	7980	7987	1	1	2	3	3	4	5	6	6
63	·7993	8000	8007	8014	8021	8028	8035	8041	8048	8055	1	1	2	3	3	4	5	5	6

Exercise 1.27

Write down the logarithms of the following numbers:

1. 3·2   2. 4·6   3. 2·31   4. 8·26
5. 9·89   6. 3·156   7. 8·209   8. 7·118
9. 1·876   10. 4·598

Logarithms of numbers greater than 10. A logarithm consists of two parts:

(1) a whole number part called the *characteristic*,
(2) a decimal part called the *mantissa*.

The characteristic depends upon the size of the number. It is found by subtracting 1 from the number of figures which occur to the left of the decimal point in the given number. Thus,

> the number 18·16 has a characteristic of 1
> the number 232·3 has a characteristic of 2
> the number 5691 has a characteristic of 3

The mantissa is found directly from the log tables.

Examples

1. To find log 7396.
   The number of figures to the left of the decimal point is 4 and hence the characteristic is 3. We now look up the log of 7·396 in the tables and find the mantissa to be 0·8690. Hence
   log 7396 = 3·8690
2. To find log 739·6.
   The characteristic is 2 and log 7·396 = 0·8690
   ∴ log 739·6 = 2·8690
4. To find log 73·96.
   The characteristic is 1 and log 7·396 = 0·8690
   ∴ log 73·96 = 1·8690
4. To find log 7·396.
   The characteristic is 0 and log 7·396 = 0·8690
   ∴ log 7·396 = 0·8690
   Numbers which have the same set of significant figures have the same mantissa in their logarithms.

Exercise 1.28

Find the logarithms of the following numbers:

1. (a) 8·3   (b) 83   (c) 830   (d) 8300
2. (a) 4·21   (b) 42·1   (c) 421   (d) 4210
3. (a) 81·92   (b) 819·2   (c) 8192   (d) 81920
4. (a) 713·6   (b) 7136   (c) 71360   (d) 713600
5. (a) 8·584   (b) 42·44   (c) 813·6   (d) 1·619

Anti-logarithms. The table of anti-logs contains the *numbers which correspond to the given logarithms.* These tables are used in a similar way to the log tables but it must be remembered that:

(1) the mantissa only of the logarithm is used when looking up anti-logs;
(2) the number of figures to the *left* of the decimal point is found by *adding 1* to the characteristic of the logarithm.

Examples

1. To find the number whose log is 3·1567. Using the mantissa 0·1567 we find in the anti-log tables that the corresponding number is 1434. Since the characteristic is 3, the number must have four figures to the left of the decimal point. The number is therefore 1434 and log 1434 = 3·1567.
2. To find the number whose log is 1·7139. From the anti-log tables we find that the number corresponding to 0·7139 is 5175. Since the characteristic is 1, the number must have two figures to the left of the decimal point. Hence the required number is 51·75.

Exercise 1.29

Find the anti-logs of the following:

1. 0·8100   2. 1·7600   3. 2·0310   4. 3·1660
5. 1·3621   6. 0·6007   7. 3·1589   8. 4·6677
9. 1·0382   10. 5·0892

Rules for the use of logarithms

1. *Multiplication.* The rule is, find the logarithms of the numbers to be multiplied and *add* them together. The answer to the multiplication is then obtained by finding the anti-log of the sum.

Example. Find 35·16 × 5·314

Number	Log	
35·16	2·5460	ADD
5·314	0·7254	

Answer = 186·8 _{anti-log} 2·2714

2. *Division.* Find the logarithm of each number and subtract the log of the denominator (bottom number) from the log of the numerator (top number). The answer to the division is obtained by finding the anti-log of the difference.

Example. Find $\dfrac{420·3}{32·16}$

Number	Log	
420·3	2·6235	SUBTRACT
32·16	1·5073	

Answer =   13·07   _{anti-log} 1·1162

Example. Find the value of $\dfrac{893·4 \times 12·02}{39·50 \times 2·761}$

Before using the log tables a *rough check* on the answer should be found. Thus,

$$\text{Rough check} = \frac{900 \times 10}{40 \times 3} = 75$$

Numerator			Denominator	
Number	Log		Number	Log
893·4	2·9511	ADD	39·50	1·5966
12·02	1·0799		2·761	0·4411
893·4×12·02	4·0310	SUBTRACT	39·50×2·761	2·0377
39·50× 2·761	2·0377			

Answer = 98·47 _{anti-log} 1·9933

Comparing the rough check and the answer there is no doubt that the decimal point is in the correct place.

Exercise 1.30
Calculate the values of the following:
1. 1·876×8·703
2. 15·15×30·68
3. 7·168×129·7
4. 2·159×900·3×11·26
5. 4·636×7·858×70·19
6. 11×2·008×90·56
7. $\dfrac{28·63}{7·164}$
8. $\dfrac{115·9}{8·390}$
9. $\dfrac{27·15×11·89}{103·5}$
10. $\dfrac{17·89×108·8×2·987}{50·63×44·68}$

Logs of numbers between 0 and 1. The characteristic of a wholly decimal number is always a negative number. Its value is numerically one more than the number of zeros which follow the decimal point. Thus,
    the number 0·7632   has a characteristic of −1
    the number 0·008 235 has a characteristic of −3

Example. To find the logarithm of 0·5617. The characteristic is −1 and the mantissa (read directly from the log tables) is 0·7495. Hence
        log 0·5617 = −1+0·7495
To write the log in this form is awkward for calculations and therefore we write
        log 0·5617 = $\bar{1}$·7495
The minus sign has been written above the characteristic but it must be clearly understood that:
        $\bar{1}$·7495 = −1+0·7495
        $\bar{3}$·2618 = −3+0·2618 and so on.
    A logarithm written in this way has a negative characteristic and a positive mantissa.

Exercise 1.31
Find the logarithms of the following:
1. (a) 30·56   (b) 3·056
   (c) 0·3056   (d) 0·003 056
2. (a) 6·352   (b) 0·6352
   (c) 0·063 52   (d) 0·006 352
3. (a) 0·7361   (b) 0·0216
   (c) 0·000 481 7   (d) 0·5968
4. (a) 0·018 23   (b) 0·009 821
   (c) 0·000 056   (d) 0·077 28

The anti-logs for logs with negative characteristics. When finding the anti-log only the mantissa is used when looking up the anti-log tables. The number of zeros following the decimal point is 1 less than the numerical value of the negative characteristic.

Example. To find the number whose log is $\bar{2}$.5231. Using the mantissa 0·5231 we find in the anti-log tables that the corresponding number is 3335. Since the characteristic is $\bar{2}$, the number must have one zero following the decimal point. The number is therefore 0·033 35 and log 0·033 35 = $\bar{2}$·5231.

Exercise 1.32
Find the anti-logs of the following:
1. $\bar{1}$·5962
2. $\bar{2}$·5962
3. $\bar{3}$·5962
4. $\bar{3}$·6059
5. $\bar{2}$·1687
6. $\bar{4}$·1108
7. $\bar{1}$·5006
8. $\bar{3}$·9210
9. $\bar{2}$·0068
10. $\bar{5}$·6212

Adding and subtracting negative characteristics. The rules are the same as when adding and subtracting directed numbers.

Examples
1. Add $\bar{2}$ and $\bar{3}$.
    $\bar{2}+\bar{3}$ = −2+(−3) = −5 = $\underline{\bar{5}}$
2. Add 3, 2 and $\bar{1}$.
    $3+2+\bar{1}$ = −3+2+(−1) = −4+2 = −2 = $\underline{\bar{2}}$
3. Add 3·5 and $\bar{2}$·2.
    3·5   The addition of the mantissa is 0·5+0·2 = 0·7.
    $\bar{2}$·2   The addition of the characteristics is 3+$\bar{2}$
    $\overline{1·7}$ = 3−2 = 1. The answer is therefore 1·7.
4. Add 1·8 and $\bar{3}$·7.
    1·8   The addition of the mantissa is 0·8+0·7 = 1·5.
    $\bar{3}$·7   The 1 is carried and the addition of the
    $\overline{\bar{1}·5}$ characteristic is 1+$\bar{3}$+1 = 2−3 = −1 = $\bar{1}$. The
            answer is therefore $\underline{\bar{1}·5}$.
5. Find $\bar{3}$−$\bar{2}$.
    $\bar{3}$−$\bar{2}$ = −3−(−2) = −3+2 = −1 = $\underline{\bar{1}}$
6. Find 2−$\bar{3}$.
    2−$\bar{3}$ = 2−(−3) = 2+3 = $\underline{5}$

7. Find $\bar{3}\cdot2-\bar{1}\cdot8$.

$\begin{array}{l}\underline{3\cdot2}\\ \bar{1}\cdot8\\ \hline \bar{3}\cdot4\end{array}$ We cannot subtract 8 from 2 and hence we borrow 1 from $\bar{3}$ making it $\bar{4}$. We now subtract 0·8 from 1·2 giving 0·4. The characteristic becomes $\bar{4}-\bar{1}=-4-(-1)=-4+1=-3=\bar{3}$. The answer is $\underline{\bar{3}\cdot4}$.

Exercise 1.33

Add the following:

1. $\bar{2}+1$
2. $3+\bar{1}$
3. $\bar{2}+3$
4. $3+\bar{1}$
5. $\bar{4}+3$
6. $2+\bar{2}$
7. $3+\bar{2}$
8. $\bar{2}+4$
9. $\bar{5}+1$
10. $0+\bar{1}$
11. $3\cdot5+\bar{2}\cdot2$
12. $\bar{2}\cdot4+\bar{2}\cdot5$
13. $2\cdot8+\bar{1}\cdot7$
14. $\bar{1}\cdot3+3\cdot9$
15. $\bar{2}\cdot5+3\cdot8$
16. $1\cdot9+\bar{1}\cdot8$
17. $\bar{1}\cdot8+0\cdot7$
18. $0\cdot8+\bar{1}\cdot3$
19. $2\cdot8+\bar{2}\cdot5$
20. $0\cdot7+\bar{2}\cdot6$

Subtract the following:

21. $4-2$
22. $3-\bar{2}$
23. $0-\bar{1}$
24. $\bar{2}-\bar{4}$
25. $3-3$
26. $\bar{2}-5$
27. $\bar{4}-\bar{2}$
28. $3-\bar{2}$
29. $\bar{4}-3$
30. $2-3$
31. $1-5$
32. $0-3$
33. $3\cdot6-2\cdot9$
34. $\bar{2}\cdot7-1\cdot6$
35. $\bar{1}\cdot5-\bar{1}\cdot2$
36. $\bar{1}\cdot8-\bar{2}\cdot6$
37. $0\cdot4-\bar{2}\cdot2$
38. $1\cdot5-1\cdot6$
39. $1\cdot4-3\cdot7$
40. $\bar{1}\cdot5-\bar{1}\cdot8$
41. $\bar{2}\cdot3-1\cdot6$
42. $\bar{1}\cdot6-2\cdot8$
43. $0\cdot8-\bar{3}\cdot9$
44. $1\cdot2-\bar{4}\cdot7$

Calculations using logarithms

1. Find the value of $0\cdot3782\times0\cdot005\,692$. Before using the log tables a rough check on the answer should be found. Thus,

Rough check $=0\cdot4\times0\cdot006=0\cdot0024$

Number	Log	
0·3782	$\bar{1}$·5777	ADD
0·005 692	$\bar{3}$·7553	

Answer = 0·002 153 anti-log $\bar{3}$·3330

Comparing the rough check and the answer we see that there is no doubt that the decimal point is in the correct position.

2. Find the value of $\dfrac{0\cdot082\,63}{0\cdot5218}$

Doing a rough check we have

Rough check $=\dfrac{0\cdot08}{0\cdot5}=0\cdot16$

Number	Log	
0·08263	$\bar{2}$·9171	SUBTRACT
0·5218	$\bar{1}$·7175	

Answer = 0·1583 anti-log $\bar{1}$·1996

The rough check and the answer compare favourably and there is no doubt that the decimal point is in the correct position.

3. Find the value of $\dfrac{0\cdot6875\times0\cdot032\,57}{17\cdot35\times0\cdot8297}$

Rough check $=\dfrac{0\cdot7\times0\cdot03}{20\times0\cdot8}=0\cdot0015$

	Numerator			Denominator	
	Number	Log		Number	Log
	0·6875	$\bar{1}$·8373	ADD	17·35	1·2392
	0·032 57	$\bar{2}$·5128		0·8297	$\bar{1}$·9190
0·6875×0·032 57		$\bar{2}$·3501	SUBTRACT	17·35×0·8297	1·1582
17·35×0·8297		1·1582			

Answer = 0·001 555 anti-log $\bar{3}$·1919

Exercise 1.34

Calculate the values of the following:

1. $0\cdot265\times4\cdot168$
2. $3\cdot672\times0\cdot976$
3. $0\cdot721\times0\cdot5932$
4. $0\cdot087\,63\times0\cdot2137$
5. $0\cdot7653\div0\cdot031\,72$
6. $30\cdot16\div0\cdot5073$
7. $0\cdot016\,32\div11\cdot16$
8. $628\cdot9\div0\cdot021\,74$
9. $0\cdot0073\div5\cdot187$
10. $0\cdot5172\times0\cdot6342\times5\cdot174$
11. $11\cdot15\times0\cdot031\,74\times0\cdot4173$
12. $\dfrac{23\cdot15\times0\cdot6325}{0\cdot1537}$
13. $\dfrac{0\cdot3173}{2\cdot153\times0\cdot8619}$
14. $\dfrac{31\cdot59\times0\cdot2517}{13\cdot15\times0\cdot0529}$
15. $\dfrac{0\cdot5219\times0\cdot6319\times0\cdot021\,73}{0\cdot4973\times0\cdot032\,15\times0\cdot038\,97}$

1.8   The slide rule

Introduction. We may add two lengths mechanically as shown in Fig. 1.19. If we wish to add lengths of 3 and 4 together we may do it by setting 0 on the upper scale opposite 3 on the lower scale. Then by finding 4 on the upper scale we find under it 7 on the lower scale. Thus the sum of 3 and 4 is 7.

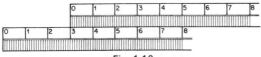

Fig. 1.19

Just as a scale of cm and mm is constructed on an ordinary scale, so it is possible to construct a rule in which the distance from the left hand to a particular number refers to the logarithm of the number.

If we remember that the log of a product is the sum of the logs of the separate numbers, then it is clear that we can use the logarithmic scales to evaluate a product. Also, if we wish to divide, we simply subtract lengths instead of adding lengths.

Construction of logarithmic scales. Suppose we wish to construct a scale of logarithms of numbers from 1 to 10 and that the total length of the scale is to be 250 mm. Since log 1 = 0 the scale will be marked 1 at the zero point. From the log tables, log 2 = 0·3010 so that the length from 1 to 2 will be 250×0·3010 = 75·25 mm. Similarly, log 3 = 0·4771 and the length from 1 to 3 will be 119·3 mm. The tabulation below shows this for each of the numbers from 1 to 10.

Number	Logarithm	Distance from zero point (mm)
1	0	0
2	0·3010	250×0·3010 = 75·25
3	0·4771	250×0·4771 = 119·3
4	0·6021	250×0·6021 = 150·5
5	0·6990	250×0·6990 = 174·8
6	0·7782	250×0·7782 = 194·6
7	0·8451	250×0·8451 = 211·3
8	0·9031	250×0·9031 = 225·8
9	0·9542	250×0·9542 = 238·6
10	1·0000	250×1 = 250

The distance between 1 and 2 is greater than that between 2 and 3 and this is greater than that between 3 and 4, the distance between 9 and 10 being least of all. We shall thus have a rule with unequal graduations. If the rule is to be any use, intermediate graduations are required, the log tables being used for the lengths between these intermediate graduations. This is the fundamental principle of the construction of a slide rule.

Fig. 1.20 shows, in a diagrammatic form, the scales on an ordinary 250 mm slide rule. It will be noticed that there are four scales, two on the rule and two on the slider. The upper and lower scales on the rule are called A and D and those on the slider B and C. The A and B scales are identical and so are the C and D scales.

Notice that on the A and B scales the figure 10 is just half way to 100. This is because log 10 = 1 and log 100 = 2. These scales really contain two short scales, that is, one scale half the length of the rule is repeated twice. The C and D scales read only from 1 to 10.

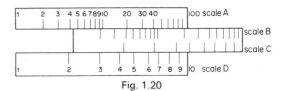

Fig. 1.20

As the distances between consecutive numbers vary, a different number of minor divisions are necessary at different points on the scales. This causes a certain amount of confusion at first but this will disappear with practice. The left hand end 1 may be called 1, 10, 100, 1000 etc., or 0·1, 0·01, 0·001, 0·0001 etc. The minor divisions always retain their relation to the index figure. Thus 757 might be read 75·7 or 0·757 etc. by supposing the index figure to be suitably altered. As we will see later on, positioning the decimal point is most important when using the slide rule.

Multiplication. As pointed out earlier, multiplication on the slide rule involves adding lengths. It is possible therefore to use the A and B scales or the C and D scales. It is more accurate to use the C and D scales because, in effect, these scales are twice as long as the A and B scales. The method is illustrated by the following examples.

Examples

1. Find 2×3 on a slide rule.

   Fig. 1.21 shows the method. Place the cursor over 2 on scale D. Bring the 1 on scale C under the cursor. Now take the cursor to 3 on scale C. Read off the answer on scale D. This is found to be 6.

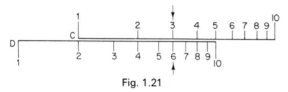

Fig. 1.21

2. Find 6×5 on a slide rule.

   As shown in Fig. 1.22 place the cursor over 6 on scale D. If we now bring the 1 on scale C under the cursor, we will find that 5 on scale C lies beyond the end of scale D. Therefore, we bring the 10 on scale C under the cursor. Now take the cursor to 5 on scale C and read off the answer on scale D. This will be 3. Now, as we pointed out earlier this 3 may be called 3, 30, 300 etc. or 0·3, 0·03, 0·003 etc. Obviously in this case the 3 represents 30 and this is the result.

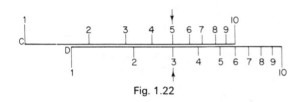

Fig. 1.22

3. Find 5·84×1·76 on a slide rule.

   Place the cursor over 5·84 on scale D. Bring the 10 on scale C under the cursor and take the cursor to 1·76 on scale C. Read off the answer on scale D. This will be seen to be 1·027. Obviously this is not the correct answer since the decimal point is incorrectly positioned. The correct position of the decimal point may easily be found by doing a rough check. Thus,

   $$5·86×1·76 \text{ is approximately } 6×2 = 12$$

   Therefore the answer is 10·27.

   Although there is another method of finding the position of the decimal point, most students will find the rough check method the better.

4. Find 0·0046×39·5×114×0·935 using a slide rule.

   Place the cursor over 4·6 on scale D. Bring the 10 on scale C under the cursor and take the cursor to 3·95 on scale C. The product of 0·0046 and 39·5 is given on scale D, *but we do not have to read this product*. Now bring the 1 on scale C under the cursor and take the cursor to 1·14. Bring the 10 on scale C under the cursor and take the cursor to 9·35 on scale C. The result is then read off on scale D and will be seen to be 1·936. To position the decimal point do a rough check.

Thus, 0·0046×39·5×114×0·935 is approximately

$$\frac{5}{1000}×40×100×1 = 20$$

The answer is, therefore, 19·36.

Division. As stated earlier, division on the slide rule involves subtracting lengths. The C and D scales are again used.

Examples

1. Find 8÷2.

   As shown in Fig. 1.23 place the cursor over 8 on scale D. Bring 2 on scale C under the cursor. Now take the cursor to 1 on scale C. Read the answer from scale D. This will be 4.

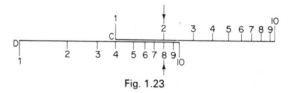

Fig. 1.23

2. Find 0·0536÷85·8.

   Place the cursor over 5·36 on scale D. Bring 8·58 on scale C under the cursor. Now take the cursor to 10 on scale C and read the answer from scale D. This will be 6·24. To position the decimal point do a rough check. Thus,

   $$\frac{0·0536}{85·8} \text{ is approximately } \frac{5}{100}×\frac{1}{80} = \frac{1}{1600} = 0·0006$$

   The answer is, therefore, 0·000 624.

2. Find $\dfrac{7·31×32·2×389}{853×0·219×38·3}$

   This example involves multiplication and division. The easiest way is:

   (1) Divide 7·31 by 853.
   (2) Multiply this quotient by 32·2.
   (3) Divide by 0·219.
   (4) Multiply this quotient by 389.
   (5) Divide by 38·3.

   The operations are performed on the slide rule as follows:

   (1) Place the cursor on 7·31 on scale D. Bring 8·53 on scale C under the cursor.
   (2) Take the cursor to 3·22 on scale C.
   (3) Bring 2·19 on scale C under the cursor.
   (4) Take the cursor to 3·89 on scale C.
   (5) Bring 3·83 on scale C under the cursor and take the cursor to 1 on scale C.
   (6) The answer is then read off from scale D, and will be seen to be 1·280.
   (7) Now do a rough check to position the decimal point.

$$\frac{7\cdot31\times32\cdot2\times389}{853\times0\cdot219\times38\cdot3}$$ is approximately

$$\frac{8\times30\times400}{800\times0\cdot2\times40} = \frac{3}{0\cdot2} = 15$$

The answer is, therefore, 12·80.

Exercise 1.35
1. Using a slide rule, perform the calculations of Exercise 1.30.
2. Using a slide rule, perform the calculations of Exercise 1.34.

1.9   Symbolic notation

Introduction. The methods of algebra are an extension of those used in arithmetic. In algebra we use symbols as well as numbers to represent quantities. When we write that a bolt has a diameter of 50 mm we are making a *particular* statement, but if we write that a bolt has a diameter of *d* mm then we are making a *general* statement. This general statement will cover any number we care to substitute for *d*.

We do not, as a rule, use multiplication signs in algebra. An expression like 5×*n* would be written simply as 5*n*. Division signs are used but it is generally better to write $\frac{a}{b}$ rather than $a \div b$. Plus and minus signs are used in the usual way, so that expressions like *x*+*y* and *a*−*b* are correct.

Use of symbols. The following examples will show how verbal statements can be translated into algebraic symbols. Notice that we can choose any symbols we like to represent the quantities concerned.

(1) The sum of two numbers.
 Let one number be *x* and the other *y*.
 Then the sum of the two numbers is *x*+*y*.
(2) Three times a number.
 Let the number be *n*.
 Then three times the number is 3×*n* or 3*n*.
(3) One number divided by another.
 Let the two numbers be *p* and *q*.
 Then one number divided by another is $\frac{p}{q}$.
(4) 8 times a number minus 6.
 Let the number be *x*.
 Then 8 times the number minus 6 is 8*x*−6.
(5) 5 times the product of two numbers.
 Let the numbers be *n* and *m*.
 Then 5 times their product is 5×*n*×*m* or 5*nm*.

Exercise 1.36
Translate the following statements into algebraic symbols:
1. 7 times a number.
2. 4 times a number minus 3.
3. 5 times a number plus 6 times a second number.
4. Sum of two numbers divided by a third number.
5. Half of a number.
6. 8 times the product of three numbers.
7. Product of two numbers divided by a third number.
8. 3 times a number minus 4 times a second number.

Substitution. The process of finding the numerical value of an algebraic expression for given values of the symbols that occur in the expression is called substitution.

Example. If $x = 3$, $y = 4$ and $z = 5$, find values for the following algebraic expressions: (a) $2x+5$ (b) $3x+2y+4z$ (c) $8-y$ (d) xyz (e) $\frac{y}{z}$ (f) $\frac{3y+z}{2z}$

(a) $2x+5 = 2\times3+5 = 6+5 = 11$
(b) $3x+2y+4z = 3\times3+2\times4+4\times5 = 9+8+20 = 37$
(c) $8-y = 8-4 = 4$
(d) $xyz = 3\times4\times5 = 60$
(e) $\frac{y}{z} = \frac{4}{5} = 0\cdot8$
(f) $\frac{3y+z}{2z} = \frac{3\times4\times5}{2\times5} = \frac{12+5}{10} = \frac{17}{10} = 1\cdot7$

Exercise 1.37
If $a = 1$, $b = 2$, and $c = 3$, find values for the following expressions:

1. $a+5$	2. $c-2$	3. $6-b$
4. $4b$	5. $5c$	6. ab
7. $3bc$	8. abc	9. $8abc$
10. $4a+3b$	11. $5c-2$	12. $a+2b+3c$

13. $8c-2b$ 14. $\frac{1}{4}b$ 15. $\frac{c}{6}$

16. $\frac{a}{3}$ 17. $\frac{ab}{8}$ 18. $\frac{c}{b}$

19. $\frac{a+2b}{c}$ 20. $\frac{3c-2}{ab}$

Powers. The quantity $a \times a \times a$ or aaa is written a^3. a^3 is called the third power of a. The small figure 3 which indicates the number of times a is multiplied by itself is called the index.

$$2^4 = 2 \times 2 \times 2 \times 2 = 16$$
$$y^5 = y \times y \times y \times y \times y$$

Example. Find the value of b^3 when $b = 3$.
$$b^3 = 3^3 = 3 \times 3 \times 3 = \underline{27}$$

When dealing with expressions like $5\,mn^3$ care must be taken. It is only the symbol n which is raised to the third power. Thus,
$$5\,mn^3 = 5 \times m \times n^3 = 5 \times m \times n \times n \times n$$

Example. Find the value of $8a^2b^3$ when $a = 2$ and $b = 3$.
$$8a^2b^3 = 8 \times 2^2 \times 3^3 = 8 \times 2 \times 2 \times 3 \times 3 \times 3 = \underline{864}$$

Exercise 1.38
If $x = 2$, $y = 3$ and $z = 4$, find values for the following expressions:

1. x^3 2. y^4 3. xy^2 4. $2x^2y$
5. xy^2z^2 6. $5x^2+3y^3$ 7. x^2+z^2 8. $7y^3z$
9. $\frac{3x^4}{z^2}$ 10. $\frac{z^5}{x^2y}$

Equations. Fig. 1.24 shows a pair of scales which are in balance, that is, the masses in the left-hand pan equal the masses in the right-hand pan.

Therefore, $x+2 = 7$

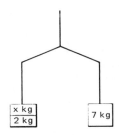

Fig. 1.24

This is an example of a simple equation. To solve this equation, we have to find a value of x such that the scales remain in balance. Now, the only way to maintain the balance of the scales is to add or subtract the same amount from each pan. If we take 2 kg from the left-hand pan, then we are left with x kg in that pan, but we must also take 2 kg from the right-hand pan, that is,
$$x+2-2 = 7-2$$
$$x = 5 \text{ kg}$$

We now take a second example, so that the scales are in balance when we have 3 exactly similar packets of sugar in the left-hand pan and a mass of 6 kg in the right-hand pan. We want to know the mass of 1 packet of sugar. If we let the mass of 1 packet of sugar be x kg, then the mass of 3 packets of sugar is $3x$ kg. Therefore, $3x = 6$.

We can maintain the balance of the scales if we multiply or divide the quantities in each pan by the same amount. If we divide each of these by 3, then we have:
$$\frac{3x}{3} = \frac{6}{3} \quad \text{i.e. } x = 2 \text{ kg}$$

From these two examples we can say:
(1) An equation expresses balance between two sets of quantities.
(2) We can add or subtract the same amount from each side of an equation.
(3) We can multiply or divide by the same amount on each side of an equation.

These rules will become clearer if you study the following examples:

(1) If $2x = 6$
$x = 3$ (divide each side by 2)

(2) If $\frac{x}{3} = 2$
$x = 6$ (multiply each side by 3)

(3) If $\frac{2x}{4} = 5$
$2x = 20$ (multiply each side by 4)
$x = 10$ (divide each side by 2)

(4) $x+4 = 9$
$x = 5$ (subtract 4 from each side)

(5) $2x+7 = 19$
$2x = 12$ (subtract 7 from each side)
$x = 6$ (divide each side by 2)

(6) $x-5 = 2$
$x = 7$ (add 5 to each side)

(7) $3x-4 = 8$
$3x = 12$ (add 4 to each side)
$x = 4$ (divide each side by 3)

(8) $\frac{6}{4} = \frac{3}{x}$
$6 = \frac{12}{x}$ (multiply each side by 4)
$6x = 12$ (multiply each side by x)
$x = 2$ (divide each side by 6)

Exercise 1.39
Solve the following equations:

1. $x+2 = 5$ 2. $x+6 = 8$ 3. $x+5 = 9$
4. $x-1 = 7$ 5. $y-3 = 2$ 6. $t-2 = 3$
7. $3x = 9$ 8. $2t = 8$ 9. $7t = 21$
10. $\frac{t}{2} = 5$ 11. $\frac{y}{3} = 7$ 12. $\frac{1}{3}x = 6$
13. $\frac{1}{4}y = 2$ 14. $2x+5 = 13$ 15. $3x+7 = 25$
16. $4x-3 = 5$ 17. $2x-6 = 8$ 18. $5z-1 = 9$

19. $\dfrac{6}{x} = 3$ 20. $\dfrac{3}{y} = 6$ 21. $\dfrac{x}{4} = \dfrac{3}{2}$

22. $\dfrac{25}{t} = 5$ 23. $\dfrac{3}{q} = \dfrac{2}{5}$ 24. $\dfrac{1}{p} = 3$

Now let us see how we can use equations to solve practical problems.

Example. Find the dimension L for the spindle shown in Fig. 1.25.

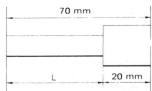

Fig. 1.25

The conditions are that the required length L together with the 20 mm dimension must equal 70 mm. Writing this statement as an equation we have,

$L+20 = 70$
$\quad L = \underline{50 \text{ mm}}$ (subtracting 20 mm from each side)

Example. Fig. 1.26 shows a bar resting in a vee block. Calculate the dimension H.

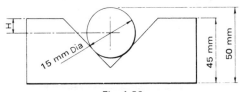

Fig. 1.26

The first step is to calculate the radius of the bar which is $\frac{15}{2} = 7.5$ mm. Then using the base of the vee block as datum we see that the dimension from the base of the block to the centre of the bar is $50-7.5 = 42.5$ mm. We also see that the dimension from the base of the block to the centre of the bar is $45-H$.

$\therefore\ 45-H = 42.5$
$\qquad H = 45-42.5 = \underline{2.5 \text{ mm}}$

Example. Rivets are spaced 40 mm apart in a plate 830 mm long. If the distance between the centres of the first and last rivets and the edge of the plate is 15 mm, find the total number of rivets required.

Let number of spaces between rivets = x
Then the total length of plate   = $40x+2\times15$ mm
But the length of the plate   = 830 mm
$\therefore\ 40x+2\times15 = 830$
$\qquad\quad 40x = 830-30$
$\qquad\quad 40x = 800$
$\qquad\qquad x = \dfrac{800}{40} = \underline{20}$.

Thus there are 20 spaces between the rivets and hence 21 rivets are required.

Exercise 1.40
The following exercises should be done by first obtaining and then solving an equation.
1. Find the dimension A shown in Fig. 1.27.
2. Find the dimension B shown in Fig. 1.28.

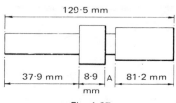

Fig. 1.27

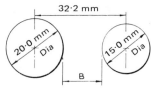

Fig. 1.28

3. Find the dimension X shown in Fig. 1.29.

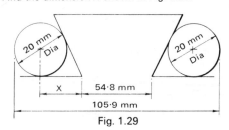

Fig. 1.29

4. A lathe tool is set to advance 0.5 mm per revolution of the spindle. Find the number of revolutions for the tool to advance 200 mm.
5. 8 equally spaced holes are required in a plate 390 mm long. If the distance between the centres of the first and last holes and the edge of the plate is to be 20 mm, find the spacing of the holes.
6. Rivets are spaced 30 mm apart in a plate 384 mm long. If the distance between the first and last rivets and the edge of the plate is 12 mm, find the number of rivets required.

Formulae. A formula is a rule which describes the relationship between two or more quantities. A formula is very similar to an equation. The statement that $E = IR$ is a formula for E in terms of I and R. The value of E may be found by simple arithmetic after substituting the given values of I and R.

Example. If $E = IR$ find the value of E when $I = 6$ and $R = 4$.

Substituting the values of I and R we have
$$E = 6 \times 4 = \underline{24}$$
Notice that in the formula we have written $E = I \times R$ without the multiplication sign, but when the figures are substituted for the symbols the multiplication sign reappears.

Example. The spindle speed for a lathe in revolutions per minute may be found from the formula
$$N = \frac{300\ S}{D}$$

where S = the cutting speed in metres per minute
and   D = the work diameter in millimetres.

Find the spindle speed for a bar 30 mm diameter if the cutting speed is to be 25 m/min. We are given $S = 25$ and $D = 30$. Substituting these values in the formula:
$$N = \frac{300 \times 25}{30} = 250$$
Hence the spindle speed required is $\underline{250\ \text{rev/min}}$.

Exercise 1.41

1. If $I = \dfrac{E}{R}$ find the value of I when $E = 20$ and $R = 4$.

2. If $v = u + at$ find the value of v when $u = 10$, $a = 3$ and $t = 2$.

3. If $P = \dfrac{RT}{V}$ find the value of P when $R = 50$, $T = 30$ and $V = 6$.

4. If $P = \dfrac{1}{n}$ find the value of P when $n = 20$.

5. If $A = \dfrac{bh}{2}$ find the value of A when $b = 80$ and $h = 50$.

6. If $V = E - IR$ find the value of V when $E = 220$, $I = 15$ and $R = 2$.

7. If $C = \pi d$ find the value of C when $\pi = \frac{22}{7}$ and $d = 140$.

8. The area of a circle is given by $A = \pi r^2$. Find A when $\pi = \frac{22}{7}$ and $r = 14$.

9. The cutting speed for turning is given by
$$S = \frac{\pi d N}{1000}$$

Where S = the cutting speed in metres per minute,
    d = the work diameter in millimetres
and N = the lathe spindle speed in rev/min.

Find S when $d = 28$ mm, $N = 200$ rev/min and $\pi = \frac{22}{7}$.

10. If $C = \dfrac{nE}{R + nr}$ find C when $n = 7$, $E = 3$, $R = 6.3$ and $r = 1.1$.

1.10 Mensuration

Units of length. In Chapter 1.3 we saw that the standard measurement of length is the metre (abbreviation: m) and that this is split up into smaller units as follows:

$$1 \text{ metre (m)} = 10 \text{ decimetres (dm)}$$
$$= 100 \text{ centimetres (cm)}$$
$$= 1000 \text{ millimetres (mm)}$$

Units of area. The area of a figure is measured by seeing how many square units it contains. A square metre is the area inside a square which has a side of 1 metre (Fig. 1.30). Similarly a square millimetre is the area inside a square which has a side of 1 millimetre.

The standard abbreviations for units of area are:

square metre = m²
square centimetre = cm²
square millimetre = mm²

Fig. 1.30

Area of a rectangle. The rectangle (Fig. 1.31) has been divided up into 4 rows of 2 squares, each square having an area of 1 square centimetre. The rectangle has, therefore, an area of 4×2 = 8 square centimetres. All we have done to find the area is to multiply the length by the breadth. The same rule will apply to any rectangle, hence,

area of rectangle = length × breadth

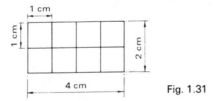

Fig. 1.31

Example. A rectangle has a length of 38 mm and a width of 25 mm. What is its area?

Area of rectangle = 38×25 = <u>950 mm²</u>

Example. A sheet of tin-plate is 420 mm long and 300 mm wide. From one corner a piece 50 mm long and 40 mm wide is removed. What area of plate remains?

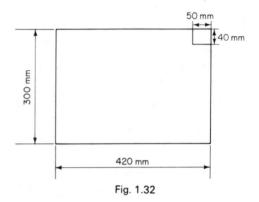

Fig. 1.32

Area of sheet = 420×300 = 126 000 mm²
Area removed = 50×40 = 2000 mm²
Area remaining = 126 000−2000 = <u>124 000 mm²</u>

Conversion of square units. A square of 1 metre side has an area of 1 square metre. It contains 100×100 = 10 000 square centimetres and it also contains 1000×1000 = 1 000 000 square millimetres.

∴ 1 m² = 10 000 cm² = 1 000 000 mm²

Example. A floor 11 m long and 7 m wide is to be covered with tiles which are 100 mm square. How many tiles are needed?

Area of floor = 11×7 = 77 m²
= 77×1 000 000 = 77 000 000 mm²
Area of tile = 100×100 = 10 000 mm²

Number of tiles needed = $\dfrac{77\,000\,000}{10\,000}$ = <u>7700</u>

Areas of figures composed of rectangles. The areas of many shapes used in engineering can be found by splitting up the shape into rectangles.

Example. Find the area of the **I**-section shown in Fig. 1.33.

The section can be divided up into the three rectangles shown.

Area of section = area of 1+area of 2+area of 3
= (50×10)+(90×8)+(150×20)
= 500+720+3000
= <u>4220 mm²</u>

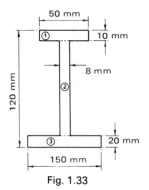

Fig. 1.33

The ratio $\dfrac{\text{Circumference}}{\text{Diameter}}$ is so important that it has been given the special symbol π (the Greek letter pi).

Circumference

Fig. 1.35

Areas of triangles

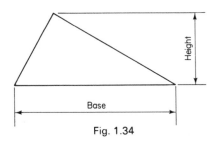

Fig. 1.34

Area of triangle = $\frac{1}{2} \times$ base $\times$ height (see Fig. 1.34)

Example. A triangle has a base 5 cm long and a vertical height of 6 cm. Find its area.

Area = $\frac{1}{2} \times 5 \times 6$ = <u>15 cm²</u>

When the lengths of the three sides of a triangle are given, its area may be found by using the formula:

Area of triangle = $\sqrt{s(s-a)(s-b)(s-c)}$

Where a, b and c are the lengths of the three sides and $s = \frac{1}{2}$ perimeter $= \frac{1}{2} \times (a+b+c)$.

Example. Find the area of a triangle whose sides are 4 cm, 6 cm and 8 cm respectively.

$s = \frac{1}{2} \times (4+6+8) = \frac{1}{2} \times 18 = 9$

Area of triangle = $\sqrt{9 \times (9-4) \times (9-6) \times (9-8)}$

$= \sqrt{9 \times 5 \times 3 \times 1}$

$= \sqrt{135}$

$= \underline{11 \cdot 62 \text{ cm}^2}$

Mensuration of the circle. The names of the main parts of a circle are shown in Fig. 1.35. The ratio

$\dfrac{\text{Circumference}}{\text{Diameter}} = 3 \cdot 14159$ ....

The exact value of this ratio has never been worked out and for most practical problems a value of $3 \cdot 142$ is sufficiently accurate when working in decimals. When working in vulgar fractions a value of $\frac{22}{7}$ can be taken.

Since $\dfrac{\text{circumference}}{\text{diameter}} = \pi$

circumference $= \pi \times diameter$

If d = diameter and r = radius, then

circumference = πd, or $2\pi r$.

Example. The diameter of a circle is 300 mm. What is its circumference?

Circumference = πd = $3 \cdot 142 \times 300$ = <u>942·6 mm</u>

Example. The radius of a circle is 14 m. What is its circumference?

Circumference = $2\pi r$ = $2 \times \frac{22}{7} \times 14$ = <u>88 m</u>

Example. A wheel 700 mm diameter makes 30 revolutions. How far does a point on the rim of the wheel travel?

In 1 revolution the point on the rim will travel a distance equal to the circumference of the wheel.

Distance travelled in 1 rev = $\pi d = \frac{22}{7} \times 700 = 2200$ mm

Distance travelled in 30 rev = 30×2200 = <u>66 000 mm or 66 m</u>

Example. A belt passes over a pulley which has a diameter of 210 mm. If the pulley rotates at 50 rev/min how far does a point on the belt travel in 1 minute?

Distance travelled by the point in 1 rev
$= \pi d = \frac{22}{7} \times 210 = 660$ mm

Distance travelled by the point in 50 rev
$= 50 \times 660$
$= 33\,000$ mm or 33 m

Since the pulley revolves 50 times in 1 minute the distance travelled by the point in 50 revolutions is <u>33 m</u>

Area of a circle. It can be shown that

Area of circle = $\pi \times (\text{radius})^2 = \pi r^2$

Example. Find the area of a circle whose radius is 30 mm.

Area = πr^2 = $3 \cdot 142 \times 30^2$ = $3 \cdot 142 \times 900$ = <u>2828 mm²</u>

Example. Find the area of a circle whose diameter is 14 cm.

Since the diameter is 14 cm, the radius $= \frac{14}{2} = 7$ cm

Area of circle = πr^2 = $\frac{22}{7} \times 7^2$ = <u>154 cm²</u>

Example. Find the area of the figure shown in Fig. 1.36. (This figure is known as an annulus.)

Area of outer circle $= \pi \times 10^2 = 314 \cdot 2$ cm²
Area of inner circle $= \pi \times 6^2 \;\; = 113 \cdot 1$ cm²
Area of annulus $\quad\;\; = 314 \cdot 2 - 113 \cdot 1 = \underline{201 \cdot 1}$ cm²

Bending allowances for sheet metal work. When a fold is made the metal at the outside of the bend stretches whilst the metal shortens at the inside of the bend. When accurate work is needed an allowance must be made to take care of this effect. The bend allowance is calculated by finding the average radius between the inner and outer radii of the bend and then by using this average radius, working out the length of arc.

Example. Find the bend allowance needed for the radius shown in Fig. 1.37.

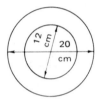

Fig. 1.36   Fig. 1.37

Outer radius $= 14$ mm
Inner radius $= 12$ mm
Average radius $= 13$ mm
To calculate the bend allowance we have to find the circumference of a quarter-circle whose radius is 13 mm.
Bend allowance $= \frac{1}{4} \times 2\pi \times 13 = \underline{20 \cdot 42}$ mm

Example. Find the width of strip needed to make the angle section shown in Fig. 1.38.

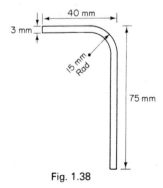

Fig. 1.38

Average radius at corner $= 13 \cdot 5$ mm
Bend allowance at corner $= \frac{1}{4} \times 2\pi \times 13 \cdot 5 = 21 \cdot 20$ mm
The width of strip needed is shown in Fig. 1.39.

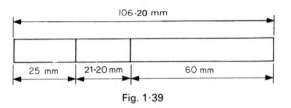

Fig. 1·39

Exercise 1.42
1. Find the areas of the following squares: (a) 4 mm side (b) 25 cm side (c) 8 m side.
2. Find the areas of the following rectangles: (a) 5 m long and 8 m wide (b) 22 cm long and 8 cm wide (c) 500 mm long and 175 mm wide.
3. Find the areas of the sections shown in Fig. 1.40.

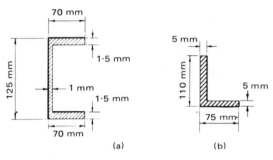

(a)   (b)

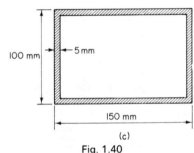

(c)

Fig. 1.40

4. A plate is 20 mm long and 15 mm wide. What is its area (a) in mm² (b) in cm²?
5. A floor is 8 m long and 10 m wide and it is to be covered with tiles which are 100 mm × 100 mm. How many tiles are needed?
6. Find the area of a triangle whose base is 125 mm and whose perpendicular height is 200 mm.
7. Find the mass of a triangular template whose base is 300 mm and whose perpendicular height is 228 mm. The metal has a mass of 30 kg/m².
8. The three sides of a triangle are 70 mm, 50 mm and 40 mm long respectively. What is its area?
9. The three sides of a triangle are 30 cm, 20 cm and 16 cm long respectively. What is its area?
10. Find the area and circumference of a circle whose radius is 35 mm. (take $\pi = \frac{22}{7}$)
11. Find the circumference and area of a circle whose diameter is 28 mm. (take $\pi = \frac{22}{7}$)

12. A pipe has an outside diameter of 42 mm and a bore of 32 mm. Find the area of its cross-section. (take $\pi = 3.142$)

13. A pulley is revolving at 700 rev/min and it has a diameter of 200 mm. How far does a point on the rim move in 1 min? $(\pi = \frac{22}{7})$

14. How many mm does a point on the rim of a wheel 140 mm diameter move when the wheel makes 30 revolutions? $(\pi = \frac{22}{7})$

15. Calculate the bend allowance needed to make each of the bends shown in Fig. 1.41.

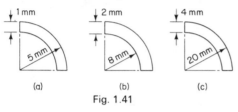

(a) (b) (c)

Fig. 1.41

16. Calculate the width of strip needed to make each of the sections shown in Fig. 1.42.

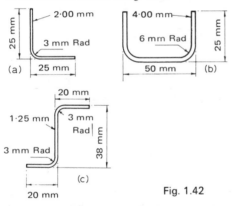

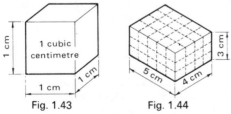

Fig. 1.42

Volume. We measure volume by seeing how many cubic units an object contains. 1 cubic centimetre (abbreviation: cm³) is the volume contained inside a cube whose edge is 1 cm (Fig. 1.43). Similarly 1 cubic metre (abbreviation: m³) is the volume contained inside a cube whose edge is 1 m.

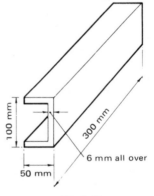

Fig. 1.43   Fig. 1.44

We can divide up the rectangular solid shown in Fig. 1.44 into 3 layers of small cubes, each cube having a volume of 1 cubic centimetre. There are 5×4 cubes in each layer and, therefore, the total number of cubes is

$5 \times 4 \times 3 = 60$. The solid, therefore, has a volume of 60 cubic centimetres. All we have done to find the volume is to multiply the length by the breadth by the height. The same rule applies to any rectangular solid.

∴ *Volume of a rectangular solid*
$$= length \times breadth \times height.$$
Since the area of the end = length × breadth we can write
Volume of a rectangular solid
$$= area\ of\ end \times height.$$

This statement is true of any solid which has a constant cross-section. For solids of this type,
$$Volume = cross\text{-}sectional\ area \times length.$$

Example. Find the volume of a rectangular block which is 32 cm long, 28 cm wide and 2 cm high.
$$Volume = 32 \times 28 \times 2 = \underline{1792\ cm^3}$$

Example. Find the volume of the channel section shown in Fig. 1.45.

Fig. 1.45

Since the bar has the same cross-section throughout its length,

Volume   = area of the end × length
Area of end = $(50 \times 6) + (50 \times 6) + (88 \times 6) = 1128\ mm^2$
Volume   = $1128 \times 300 = \underline{338\,400\ mm^3}$

Volume of a cylinder. A cylinder (Fig. 1.46) has the same section throughout its length. Hence,
Volume = cross-sectional area × length
$$= \pi r^2 h$$

Fig. 1.46

The area of the curved surface is:
Area = πdh

That is, the area of the curved surface is the circumference of the cross-section multiplied by the length (or height) of the cylinder.

Example. A pipe has the dimensions shown in Fig. 1.47. Find its volume.

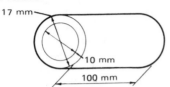

17 mm
10 mm
100 mm   Fig. 1.47

Volume	= cross-sectional area × length
Cross-sectional area	$= \pi \times 8 \cdot 5^2 - \pi \times 5^2 = 148 \cdot 37$ mm²
Volume of pipe	$= 148 \cdot 37 \times 100 = \underline{14\,837}$ mm³

Example. Find the total surface area of a cylinder whose diameter is 28 cm and whose length is 50 cm.

The total surface area is composed of the areas of the two ends plus the area of the curved surface.

Area of one end	$= \pi r^2 = \frac{22}{7} \times 14 \times 14 = 616$ cm²
Area of curved surface	$= \pi dh = \frac{22}{7} \times 28 \times 50 = 4400$ cm²
Total surface area	$= 2 \times 616 + 4400 = \underline{5632}$ cm²

Conversion of units of volume. A cube of 1 m side has a volume of 1 m³ and it contains $100 \times 100 \times 100$ = 1 000 000 cm³

1 cubic metre also contains $1000 \times 1000 \times 1000$
= 1 000 000 000 mm³
1 m³ = 1 000 000 cm³ = 1 000 000 000 mm³

Example. Convert 1·3 m³ into cm³.
$$1 \text{ m}^3 = 1\,000\,000 \text{ cm}^3$$
$$\therefore 1 \cdot 3 \text{ m}^3 = 1 \cdot 3 \times 1\,000\,000 = \underline{1\,300\,000 \text{ cm}^3}$$

Example. Convert 80 000 mm³ into cm³.
$$1000 \text{ mm}^3 = 1 \text{ cm}^3$$
$$\therefore 80\,000 \text{ mm}^3 = \frac{80\,000}{1000} = \underline{80 \text{ cm}^3}$$

Units of capacity. The capacity of a container is the volume that it will hold. It is sometimes measured in the same way as volume, that is, in cubic metres, cubic centimetres or cubic millimetres.

However, for liquid measure the litre (abbreviation: l) is frequently used. The litre is not a precise measurement since 1 litre is actually 1000·028 cubic centimetres, but for most practical purposes the litre can be assumed to be 1000 cubic centimetres.

Example. A rectangular tank has inside measurements of 2 m×1·5 m×3 m. When full, how many litres will it hold?

Volume	$= 2 \times 1 \cdot 5 \times 3 = 9$ m³
	$= 9 \times 1\,000\,000 = 9\,000\,000$ cm³

Since 1 litre = 1000 cm³

Capacity $= \frac{9\,000\,000}{1000} = \underline{9000 \text{ litres}}$

Exercise 1.43
1. Find the volume of a rectangular block 80 mm long, 50 mm wide and 35 mm high.
2. Fig. 1.48 shows the cross-section of an angle bar which is 250 mm long. Calculate its volume.

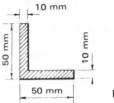

10 mm
50 mm
10 mm
50 mm   Fig. 1.48

3. Find the volume of a cylinder whose radius is 70 mm and whose height is 500 mm. ($\pi = \frac{22}{7}$)
4. A hole 40 mm diameter is drilled in a plate 25 mm thick. What volume of metal is removed. ($\pi = 3 \cdot 142$)
5. A rectangular plate 75 mm×50 mm×50 mm deep has a 13 mm diameter hole drilled right through it. What is the volume of the drilled plate? ($\pi = 3 \cdot 142$)
6. A machined part 80 mm long has the cross-section shown in Fig. 1.49. What is its volume?

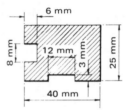

6 mm
8 mm
12 mm
3 mm
25 mm
40 mm   Fig. 1.49

7. A rectangular tank is 3 m long, 2 m wide and 1·5 m high internally. How many litres does it hold when full?
8. A cylindrical tank has inside dimensions which are: 400 mm radius and 1200 mm high. How many litres does it hold when full?
9. A triangular trough has a uniform cross-section which is a triangle having sides of 6 cm, 5 cm and 5 cm respectively. If the trough is 50 cm long, what is its volume?
10. A pipe has a 75 mm outside diameter and a 60 mm inside diameter. If it is 2 m long, what is its volume?
11. Find the total surface area of a cylinder 14 cm diameter and 20 cm long. (take $\pi = \frac{22}{7}$)
12. Find the curved surface of a cylinder which is 30 cm long and 15 cm in diameter. (take $\pi = 3 \cdot 142$)

Units of mass. The mass of an object is the amount of matter that the object contains. When we use scales we are comparing the mass of an object with a standard mass. The standard unit of mass is the kilogram (abbreviation: kg). For very small masses the gram (g) is used and

1000 grams = 1 kilogram

Density and mass. The density of a substance is the mass per unit volume. Densities are usually measured in kilogrammes per cubic metre (kg/m³). Steel, for instance, has a density of 7750 kg/m³. The mass of an object may be found by using the formula:

mass = density × volume of material in the object

DENSITIES OF METALS

Metal	Density (kg/m³)	Metal	Density (kg/m³)
Lead	11 350	Steel	7 750
Copper	8 860	Cast iron	7 200
Brass	8 580	Zinc	7 200
Tin	7 200	Duralumin	2 770
Aluminium	2 690		

Example. A cylindrical bar of copper is 80 cm long and 4 cm in diameter. What is the mass of the bar?

Volume of the bar $= \pi \times 2^2 \times 80 = 1005$ cm³

$$= \frac{1005}{1\,000\,000} = 0.001\,005 \text{ m}^3$$

From the table above, the density of copper is 8860 kg/m³.

∴ Mass of bar = Volume × density
= 0.001 005 × 8860 = <u>8·904 kg</u>

Example. Fig. 1.50 shows the cross-section of a cast iron bar which is 400 mm long. What is the mass of the bar?

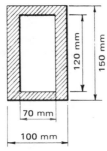

Fig. 1.50

Area of cross-section $= 150 \times 100 - 120 \times 70$
$= 6600$ mm²

Volume of bar $=$ length × area of cross-section
$= 400 \times 6600$
$= 2\,640\,000$ mm³

The density of cast iron is 7200 kg/m³

∴ Mass of bar $= \dfrac{2\,640\,000}{1\,000\,000\,000} \times 7200$

$= $ <u>19·008 kg</u>

Multiples and sub-multiples of the basic units.
For many applications some of the basic units are too small or too large and hence multiples and sub-multiples

are often needed. These are given special names as follows:

Multiplication factor	Prefix	Symbol
1 000 000 000 000	tera	T
1 000 000 000	giga	G
1 000 000	mega	M
1 000	kilo	k
100	hecto	h
10	deca	da
0·1	deci	d
0·01	centi	c
0·001	milli	m
0·000 001	micro	μ

Where possible multiples and sub-multiples should be written in powers of 1000. Thus 5000 metres should be written as 5 km not 50 hm.

Double prefixes are not permitted. For example 1000 km <u>cannot</u> be written as 1 k km but only as 1 Mm. Again 0·000 006 km cannot be written as 6 μkm but only as 6 mm.

Examples

1. Express 203 560 kg as the highest multiple possible.
   203 560 kg = 203 560 000 grams
   $= 203 \cdot 56 \times 1\,000\,000$ grams
   $= 203 \cdot 56$ megagram
   $= $ <u>203·56 Mg</u>

   (It is preferable to use 203·56 Mg rather than 0·203 56 Gg)

2. A measurement is taken as 0·000 082 m. Express this measurement as a standard multiple of a metre.
   0·000 082 $= 82 \times 0 \cdot 000\,001$ m
   $= $ <u>82 μm</u>

Exercise 1.44

1. Find the mass of a copper bar 35 mm diameter and 1000 mm long. The density of copper is 8860 kg/m³. (take $\pi = 3 \cdot 142$)

2. Find the mass of the I-section shown in Fig. 1.51 if its length is 1·2 m. The density of the metal is 7750 kg/m³.

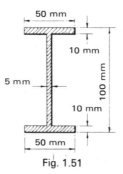

Fig. 1.51

3. A pipe 3 m long has an outside diameter of 60 mm and a bore of 50 mm. Find its mass if the density of the metal is 7750 kg/m³. ($\pi = 3\cdot142$)
4. An iron casting is shown in Fig. 1.52. What is its mass if the density of iron is 7200 kg/m³.

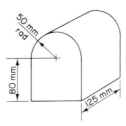

Fig. 1.52

5. A cylindrical tank 1·5 m diameter and 2 m high internally is filled with water. If the density of water is 1000 kg/m³, what mass of water does it hold? ($\pi = 3\cdot142$)
6. A bar of steel is turned down from 100 mm diameter to 90 mm diameter over a length of 200 mm. If steel has a density of 7750 kg/m³ what mass of metal is removed?
7. 3 holes each 12 mm diameter are drilled in an iron casting. What mass of metal is removed if the holes are each 50 mm long. (density of cast iron =7200 kg/m³)
8. Express each of the following as a multiple or submultiple using the standard symbols:
    (a) 8000 m   (b) 15 000 kg   (c) 3800 km
    (d) 1 891 000 kg   (e) 0·007 m   (f) 0·028 kg
    (g) 0·000 36 km   (h) 0·0036 A

1.11   Speeds and feeds

Speed. Speed is the distance travelled in unit time. For instance, a motorist might travel 65 km in 1 hour. His speed is 65 kilometres per hour. To calculate speed we use the formula:

$$speed = \frac{distance\ travelled}{time\ taken}$$

The unit of speed depends upon the units used for distance and time. If the distance travelled is measured in kilometres and the time is measured in hours then the speed is measured in kilometres per hour (km/h). If the distance travelled is measured in metres and the time is measured in minutes then the speed is measured in metres per minute (m/min).

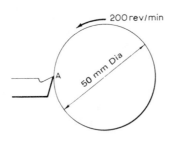

Fig. 1.53

Cutting speed for turning. When work is being turned, the cutting speed is the speed at which the material passes over the tool point. Fig. 1.53 shows a 50 mm diameter bar being machined in a lathe. The bar revolves at 200 rev/min and we wish to find the cutting speed.

If we can find the distance moved in 1 minute by a point such as A on the surface of the bar, then we shall have found the cutting speed.

Circumference of bar = 50×π 157·10 mm

$$= \frac{157\cdot10}{1000} = 0\cdot1571\ m$$

Distance moved by point A in 200 revolutions
$$= 200\times0\cdot1571 = 31\cdot42\ m$$

Since the bar revolves 200 times in 1 minute, this is the distance moved by the point A in 1 minute.

∴ Cutting speed = 31·42 m/min

From this example it can be seen that:

Cutting speed
$$= \frac{\pi\times work\ dia\ (mm)\times spindle\ speed\ (rev/min)}{1000}$$

If S = cutting speed (m/min)
        d = work diameter (mm)
and N = spindle speed (rev/min)

Then $S = \dfrac{\pi dN}{1000}$

If a rough approximation of the cutting speed is required then it is good enough to take

$$S = \frac{dN}{300}$$

The cutting speed used depends upon the tool material, the material being cut and some other factors. The table below will give some idea of the cutting speeds used when turning.

TABLE OF CUTTING SPEEDS

Metal being machined	Using a carbon steel tool	Using a high-speed steel tool
Cast iron	18 m/min	27 m/min
Mild steel	18 m/min	27 m/min
Tool steel	10 m/min	15 m/min
Brass	33 m/min	50 m/min
Aluminium	200 m/min	300 m/min

Example. A 75 mm diameter bar is being machined in a lathe at 150 rev/min. Find the cutting speed.

We are given d = 75 mm and N = 150 rev/min and we have to find S.

$$S = \frac{\pi dN}{1000} = \frac{\pi \times 75 \times 150}{1000} = \underline{35 \text{ m/min}}$$

The term 'surface speed' is often used instead of cutting speed. The formula for the cutting speed used when turning may also be used to find the surface speed of a grinding wheel and for finding the speed of a belt passing over a pulley.

Example. Calculate the speed of a belt passing over a pulley which is 500 mm diameter and which rotates at 100 rev/min.

$$S = \frac{\pi dN}{1000} = \frac{\pi \times 500 \times 100}{1000} = \underline{157 \text{ m/min}}$$

Finding the spindle speed. Suppose we have to turn a brass bar whose diameter is 75 mm at a cutting speed of 30 m/min. What spindle speed do we set on the lathe? Problems of this kind are always occurring in the machine workshop and we can find the spindle speed by using the formula:

$$N = \frac{1000 \, S}{\pi d}$$

This formula will give an accurate value for the spindle speed, but it is often good enough to use the approximate formula:

$$N = \frac{300 \, S}{d}$$

Example. A mild steel bar 50 mm diameter is to be turned using a high-speed steel tool. The spindle speeds available on the lathe are 45, 80, 110, 190, 250, 400, 600 and 1000 rev/min. Which spindle speed should be chosen?

From the table, the cutting speed for mild steel is 27 m/min.

$$N = \frac{1000 \, S}{\pi d} = \frac{1000 \times 27}{\pi \times 50} = 172 \text{ rev/min}$$

It is safer to take the nearest spindle speed below that calculated and hence *a speed of 110 rev/min is suitable.*

Example. A 75 mm diameter cast iron bar is to be turned at a cutting speed of 25 m/min. Find a suitable spindle speed.

Using the approximate formula,

$$N = \frac{300 \, S}{d} = \frac{300 \times 25}{75} = \underline{100 \text{ rev/min}}$$

Cutting speeds for milling and drilling machines. With both milling and drilling machines it is the cutter that revolves and not the work. *The cutting speed is the speed at which the tool moves over the work.* To find the cutting speed we use the formula:

$$S = \frac{\pi dN}{1000}$$

where d = the cutter diameter in millimetres.

Example. A 12 mm diameter drill revolves at 400 rev/min. Find the cutting speed.

$$S = \frac{\pi dN}{1000} = \frac{\pi \times 12 \times 400}{1000} = \underline{15 \text{ m/min}}$$

Example. A job has to be milled at a cutting speed of 20 m/min. If a 75 mm diameter cutter is used calculate the revolutions per minute of the cutter.

$$N = \frac{1000 \, S}{\pi d} = \frac{1000 \times 20}{\pi \times 75} = \underline{85 \text{ rev/min}}$$

Feeds on a lathe. When a cylindrical bar is being machined in a lathe, the tool is fed along the bar. This gives the effect of removing the metal by turning a very fine spiral. *The feed of the tool is the distance it moves for each revolution of the workpiece.*

Feeds are expressed as so many millimetres for each revolution of the workpiece (e.g. 0·5 millimetres per revolution).

Example. A lathe is set to give a feed of 0·25 mm/rev. Find how far the tool will move in 100 revolutions of the workpiece.

Distance moved by tool = feed × number of revolutions
= 0·25 × 100
= 25 mm

Knowing the feed and the cutting speed we can calculate the time to take one cut along the work.

Example. A cylindrical job 100 mm diameter is to be turned at a cutting speed of 25 m/min, the feed being 1·5 mm/rev. If the length of the job is 150 mm find the time required for 1 cut.

Step 1. Find the rev/min of the workpiece.

$$N = \frac{1000 \, S}{\pi d} = \frac{1000 \times 25}{\pi \times 100} = 80 \text{ rev/min}$$

Step 2. Find the number of revolutions to cut a length of 150 mm.

$$\text{Number of revolutions required} = \frac{\text{length of work}}{\text{feed per revolution}}$$

$$= \frac{150}{1\cdot5} = 100 \text{ rev}$$

Step 3. Time for 1 cut $= \dfrac{\text{revolutions required}}{\text{rev/min of workpiece}}$

$$= \tfrac{100}{80} = 1\cdot25 \text{ min}$$

Feeds on a drilling machine. The feed on a drilling machine is the distance the drill penetrates the work for each revolution of the drill. Feeds are usually stated in millimetres per revolution (e.g. 0·10 mm/rev). Times for drilling holes may be calculated in a similar way to that used for lathes.

Example. Calculate the time needed to drill a 10 mm diameter hole through a plate 50 mm thick. The cutting speed is to be 15 m/min and the feed 0·20 mm/rev.

Speed of drill is given by:

$$N = \frac{1000\,S}{\pi d} = \frac{1000 \times 15}{\pi \times 10} = 480 \text{ rev/min}$$

$$\text{Revolutions required} = \frac{\text{length of hole}}{\text{feed per revolution}}$$

$$= \frac{50}{0\cdot20} = 250 \text{ rev}$$

Time needed $= \dfrac{\text{revolutions required}}{\text{rev/min of drill}}$

$$= \frac{250}{480} = 0\cdot52 \text{ min}$$

Feeds on a milling machine. Feeds for milling cutters are usually expressed in millimetres per tooth (e.g. 0·10 mm per tooth). Feeds are expressed in this way because they give an indication of the amount of work each tooth is doing.

Feed per revolution = number of teeth × feed per tooth

On some milling machines the feed has to be set in millimetres per minute and this can be calculated as follows:

Feed in millimetres per minute
   = cutter rev/min × feed per rev

Example. A milling cutter has 12 teeth and a feed of 0·12 mm per tooth is to be used. If the cutter makes 90 rev/min calculate: (a) the feed in millimetres per revolution, (b) the feed in millimetres per minute.

(a) Feed in millimetres per revolution
   = number of teeth × feed per tooth
   = 12 × 0·12 = 1·44 mm/rev

(b) Feed in millimetres per minute
   = cutters rev/min × feed per rev
   = 90 × 1·44 = 12·96 mm/min

Exercise 1.45

1. A bar of steel 50 mm diameter is being turned at 80 rev/min. What is the cutting speed?
2. A metal bar 250 mm diameter is being turned at 40 rev/min. What is the cutting speed?
3. A 60 mm brass bar is to be turned at a cutting speed of 50 m/min. What spindle speed should be used?
4. A mild steel bar 75 mm diameter is to be turned at 20 m/min. The choice of machine speeds is as follows: 20, 30, 45, 68, 100, 150 and 230 rev/min. What machine rev/min should be used?
5. A 5 mm diameter hole is to be drilled in an aluminium casting. If the cutting speed is to be 30 m/min, calculate a suitable spindle speed.
6. A 15 mm diameter hole is to be drilled in cast iron at a cutting speed of 30 m/min. The spindle speeds available are: 250, 320, 450, 680, 850 and 1050 rev/min. What machine speed should be used?
7. When turning a bar 200 mm long the machine speed is 200 rev/min. If the feed is 0·10 mm/rev find the time taken to take one cut along the bar.
8. A 5 mm diameter hole is to be drilled in 25 mm thick plate. The cutting speed is to be 20 m/min and the feed 0·05 mm/rev. Find the time needed to drill the hole.
9. A cylindrical milling cutter is 60 mm diameter. It is to cut at 30 m/min. Find the rev/min of the cutter.
10. A workpiece is to have a surface speed of 30 m/min. If the work is 60 mm diameter, calculate the rev/min required.
11. Calculate the speed of a belt in metres per minute passing over a 450 mm diameter pulley rotating at 80 rev/min.
12. The table feed for a milling machine is 100 mm/min. If the length of a job is 400 mm, calculate the time needed for 1 cut.
13. The following relates to a milling machine: cutter feed = 0·10 mm per tooth; cutting speed = 40 m/min; cutter diameter = 150 mm; number of teeth in cutter = 12; cutter travel = 250 mm. Calculate the time taken for 1 cut.
14. A cylindrical milling cutter is 80 mm diameter and it has 14 teeth. It is to cut mild steel at a cutting speed of 30 m/min with a feed of 0·10 mm per tooth. Calculate (a) the cutter rev/min; (b) the table feed in mm/min; (c) the time taken for the cutter to travel 200 mm.
15. The available speeds on a lathe are: 80, 110, 190, 250, 400, 600 and 1000 rev/min. Choose the best spindle speed from the above to give a cutting speed of 25 m/min when cutting a bar 50 mm diameter.

1.12   Graphs

Axes of reference. In drawing a graph we first draw two lines at right angles to each other (Fig. 1.54). These lines are called the *axes of reference*. Their intersection (the point O) is called the *origin*.

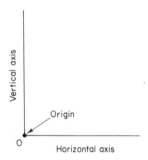

Fig. 1.54   Axes of reference

Scales. The number of units represented by a unit length measured along an axis is called the *scale*. Thus 1 cm could represent 10 units. The scales need not be the same on both axes.

Co-ordinates. Co-ordinates are used to mark the points on the graph. In Fig. 1.55 values of x are plotted against values of y. The point P has been plotted so that x = 6 and y = 15. The values of 6 and 15 are said to be the *rectangular co-ordinates* of the point P.

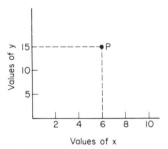

Fig. 1.55   Rectangular co-ordinates

Plotting a graph

Example. Masses per square metre of steel sheet corresponding to various thicknesses of sheet are given in the table below. Plot this information and from it find:

(a) the mass of a sheet 0·08 mm thick,
(b) the thickness of sheet corresponding to a mass of 1 kg/m².

Thickness (mm)	0·05	0·10	0·15	0·20	0·25
Mass (kg/m²)	0·4	0·8	1·2	1·6	2·0

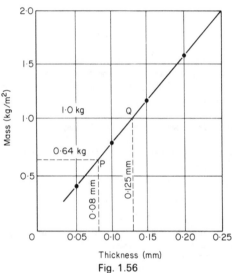

Fig. 1.56

The graph is shown plotted in Fig. 1.56 and it will be seen that it is a straight line. To find the mass of the sheet corresponding to a thickness of 0·08 mm we find 0·08 on the horizontal axis and draw a vertical line to meet the graph at P. From P draw a horizontal line to meet the vertical axis. We read off this value which is 0·64. Thus the mass of a sheet 0·08 mm thick is 0·64 kg/m².

To find the thickness of sheet corresponding to a mass of 1 kg/m² find 1 on the vertical axis and draw a horizontal line to meet the graph at Q. From Q draw a vertical line to meet the horizontal axis. Read off this value which is 0·125. Thus a sheet with a mass of 1 kg/m² has a thickness of 0·125 mm.

Example. The figures in the table below give the cutting-tool life (t minutes) between regrinds for various cutting speeds (S metres per minute).

t (min)	60	90	120	150	180
S (m/min)	33·6	28·3	27·0	26·0	25·3

Draw a graph of this information and find the tool life when the cutting speed is 30 m/min.

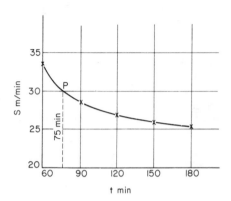

Fig. 1.57

The graph is shown in Fig. 1.57 and it is seen to be a curve. It will be noticed that on the horizontal axis the values start at 60 and on the vertical axis at 20. By doing this we are able to use larger scales to represent the quantities and hence a more accurate graph results.

To find the tool life when the cutting speed is 30 m/min we find 30 on the vertical axis and draw a horizontal line to meet the graph at P. From P we draw a vertical line to meet the horizontal axis and read off this value. It is found to be 75 and hence the tool life is 75 min when the cutting speed is 30 m/min.

Exercise 1.46

1. The distance across flats for metric hexagonal nuts of various diameters is shown in the table below. Plot a graph of this information.

Dia of bolt (mm)	0·5	1·0	1·5	2·0	2·5	3·0
Distance across flats (mm)	5·5	10·0	17·0	24·0	32·0	41·0

Plot the bolt diameter horizontally and the distance across flats vertically.

2. The circumference of circles of various diameters is shown in the table below.

Dia of circle (mm)	1	2	3	4	5
Circumference (mm)	3·14	6·28	9·4	12·6	15·7

Plot the diameter horizontally and the circumference vertically. Find, from the graph, the circumference of a circle whose diameter is 3·25 mm.

3. The spindle speeds for various drills when drilling tool steel are shown in the table below.

Drill dia (mm)	3	6	12	20	40
Drill speed (rev/min)	1000	500	250	150	75

Plot the drill diameter horizontally and the drill speed vertically. From the graph find the drill speed for drilling a 15 mm diameter hole.

4. The table below gives the efficiency of a lifting machine when it is lifting various loads.

Load (kg)	8	16	20	24	28	32
Efficiency (%)	67	83	88	90	91	92

Plot a graph of the above table with the load on the horizontal axis. From the graph find the efficiency under a load of 22 kg.

5. The table below shows corresponding values for the load and effort of a simple machine. Plot a graph of effort against load with the effort on the vertical axis. From the graph find the load corresponding to an effort of 2 kg and the effort required to lift a load of 12 kg.

Load (kg)	0	4	8	16	24
Effort (kg)	0·8	1·6	2·4	4·0	5·6

6. If the cutting speed of round bar of steel being turned in a lathe is to be 30 m/min then the rev/min needed for various diameters is given in the table:

Dia of bar (mm)	5	6	8	10	15	25
Rev/min	2000	1700	1250	1000	670	400

Plot a graph with diameter of bar on the horizontal axis. From the graph find the rev/min required for a 12 mm diameter bar.

7. In order to establish the best rake angle for a lathe roughing tool, tests were made to find how much power (in watts) was used when cutting with tools having various rake angles. The following readings were obtained:

Rake angle (degrees)	0	10	15	20	30	40
Power used (watts)	1575	1035	880	790	700	675

Plot a graph of this information with the rake angle horizontal. Find the power used when the rake angle is 25°.

1.13 Angles

When two lines meet at a point they form an angle. The size of an angle depends only on the amount of opening between the two lines. It does not depend on the length of the lines forming the angle. In Fig. 1.58 the angle A is larger than the angle B even though the lengths of its arms are shorter.

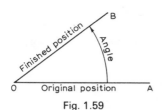

Fig. 1.58

Angular measurement. An angle may be looked upon as an amount of rotation or turning. In Fig. 1.59 the line OA has been turned about O until it takes up the position

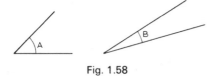

Fig. 1.59

OB. The angle through which the line has turned is the amount of opening between the lines OA and OB. If the line OA is rotated until it returns to its original position it will have described one revolution. We can therefore measure an angle as *a fraction of a revolution*. Fig. 1.60 shows a circle divided up into 36 equal parts. The first division is split up into 10 equal parts so that each small division is $\frac{1}{360}$ of a complete revolution. We call this smallest division a *degree*.

1 degree = $\frac{1}{360}$ part of a complete revolution

or, 360 degrees = 1 complete revolution.

When writing angles we write 90 degrees as 90°. The small ° at the right-hand corner of the figure replaces the word degrees. Thus, 57° reads 57 degrees.

Although the degree is a very small unit it is not small enough for some work and it is sub-divided into minutes and seconds as shown in the following table.

Angular measurement
60 seconds = 1 minute
60 minutes = 1 degree
360 degrees = 1 revolution

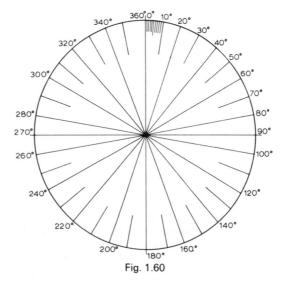

Fig. 1.60

An angle of 25 degrees 7 minutes 30 seconds is written 25° 7′ 30″. The minutes are indicated by a single tick and the seconds by a double tick.

The *right-angle* is $\frac{1}{4}$ of a revolution and a right-angle contains $\frac{1}{4} \times 360°$ = 90°. Two right-angles contain 180° and three right-angles 270°. Referring to Fig. 1.61 you will see that an *acute angle* is smaller than a right-angle. An *obtuse angle* lies between 90° and 180° whilst a *reflex angle* is greater than 180°.

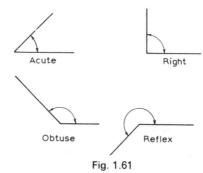

Fig. 1.61

Measurement of angles. To measure and set out angles a *protractor* is used. This instrument may be either semi-circular (the most common) or circular. It may be made from celluloid, brass, steel or wood. Fig. 1.62 shows a semi-circular protractor, the marks being 1° apart.

. . In the workshop, angles are measured by some form of protractor. The instruments used are the bevel, the combination set and the vernier protractor.

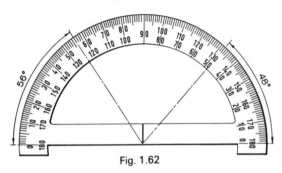

Fig. 1.62

EXAMPLES IN THE USE OF ANGLES

1. Find the angle in degrees corresponding to:
   (a) $\frac{1}{5}$ of a revolution (b) 0·75 of a revolution
   (a) 1 revolution = 360°
   ∴ $\frac{1}{5}$ revolution = $\frac{1}{5}$×360° = <u>72°</u>
   (b) 0·75 revolution = 0·75×360° = <u>270°</u>

2. In the circular plate shown in Fig. 1.63 a sector of 60° is cut out and the remainder divided up to give 5 holes equally spaced. Find the angle A between the holes.

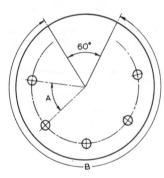

Fig. 1.63

Angle B = 360°−60° = 300°
number of equal angles = 6
∴ Angle A = $\frac{300}{6}$ = <u>50°</u>

3. Add together (a) 37°, 18° and 5° (b) 25° 39′, 48° 43′ and 70° 25′ (c) 6° 22′ 19″ and 23° 17′ 43″.
   (a)   37°+18°+5° = <u>60°</u>

   (b)   25° 39′   The minutes 39, 43 and 25 add up
       48° 43′   to 107′ since 60′ = 1°, 107′ =
       <u>70° 25′</u>   1° 47′. The 47′ is written in the
       144° 47′   minutes column and the 1° car-
                 ried over to the degrees column.
                 25°+48°+70°+1° = 144°.

   (c)   16° 22′ 19″   The seconds 19 and 43 add up to
       23° 17′ 43″   62″. Since 60″ = 1′, 62″ = 1′ 2″.
       39° 40′ 2″   The 2″ is written down in the
                    seconds column and the 1′ car-
                    ried over to the minutes column.
                    22′+17′+1′ = 40′. The 40 is
                    written down in the minutes
                    column. Since there is no carry-
                    over from the minutes to the
                    degree column, 16°+23° = 39°.

4. Subtract 29° 47′ from 48° 19′.
       48° 19′   We cannot subtract 47′ from 19′
       29° 47′   so we borrow 1° from the 48°.
       18° 32′   Since 1° = 60′ we add 60′ to 19′
                making 79′. Subtracting 47′ from
                79′ gives 32′. As we have bor-
                rowed 1° from 48°, this becomes
                47°. Subtracting 29° from 47°
                gives 18°.

Sometimes angles are quoted in decimals of degrees, for instance 37·3°, 0·57° etc. We therefore often need to convert from minutes and seconds to decimals of a degree and vice versa.

Example. Convert 15 minutes to a decimal of a degree.
   Since 1° = 60′
       15′ = $\frac{15}{60}$ = <u>0·25°</u>

Example. Convert 31′ 24″ to a decimal of a degree.
   Since 60″ = 1′
       24″ = $\frac{24}{60}$ = 0·4′
   ∴   31′ 24′ = 31·4′ = $\frac{31·4}{60}$ = <u>0·523°</u>

Example. Convert 0·3° to minutes.
   Since   1° = 60′
       0·3° = 60×0·3 = <u>18′</u>

Example. Convert 0·28° to minutes and seconds.
   0·28° = 0·28×60 = 16·8′
   0·8′ = 0·8 ×60 = 48″
   ∴ 0·28° = <u>16′ 48″</u>

Angle blocks. Angle blocks are used in the workshop for checking angles. A set may consist of the following:
   41°, 27°, 9°, 3° and 1°.

This set, when used in various combinations, will give angles up to 90° in steps of 1°. Angles may be added or subtracted by using the appropriate blocks.

Example. Make up angle blocks to give the following angles: (a) 32° (b) 50° (c) 19° (d) 43° (e) 59° (f) 5°.
   (a) 32° = 41°−9°   (d) 43° = 41°+3°−1°
   (b) 50° = 41°+9°   (e) 59° = 41°+27°−9°
   (c) 19° = 27°−9°+1°   (f) 5° = 9°−3°−1°

The blocks are shown arranged in Fig. 1.64.

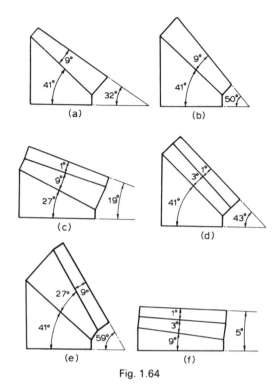

Fig. 1.64

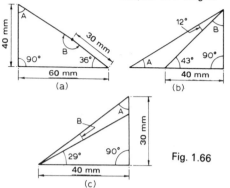

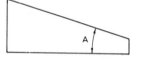

Fig. 1.66

(c) 41° 39′ and 37° 46′ (d) 11° 22′, 19° 41′ and 27° 58′ (e) 8° 51′, 7° 48′, 39° 59′, 44° 53′ and 5° 44′.

7. Add together (a) 10° 3′ 15″ and 8° 4′ 30″ (b) 18° 9′ 39″ and 15° 8′ 43″ (c) 42° 46′ 39″ and 53° 45′ 49″ (d) 18° 43′ 24″, 7° 18′ 9″ and 60° 0′ 38″.

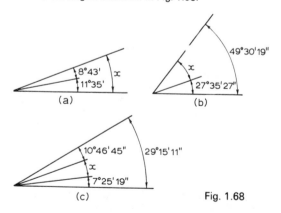

Fig. 1.67

8. Subtract (a) 12° 32′ from 19° 37′ (b) 25° 48′ from 32° 6′ (c) 15° 31′ 19″ from 18° 17′ 12″.

9. You are supplied with three angle blocks of the type shown in Fig. 1.67, with the angle A = 27°, 9° and 3° respectively. Show by means of dimensional sketches how to arrange *two or more* of these blocks to give an angle of (a) 18° (b) 30° (c) 39° (d) 33° (e) 21° (f) 12°.

10. Find the angles marked *x* in Fig. 1.68.

Example. Two angle blocks are available, one with an angle A = 25° and the other with an angle B = 48°. State the angles C and D when they are arranged as shown in Fig. 1.65.

C = 25+48 = <u>73°</u>
D = 48−25 = <u>23°</u>

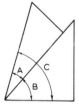

Fig. 1.65

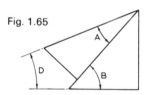

Exercise 1.47

1. Using an ordinary celluloid protractor draw the following angles: (a) 27° (b) 45° (c) 78° (d) 112°.
2. Using a protractor and rule draw the diagrams shown in Fig. 1.66. In each case measure the angles A and B.
3. Find the angle in degrees corresponding to the following fractions of a revolution: (a) $\frac{1}{20}$ (b) $\frac{1}{8}$ (c) 0·25 (d) $\frac{1}{3}$ (e) 0·3.
4. How many degrees are there in (a) 2 right-angles (b) $\frac{3}{5}$ of a right-angle (c) $\frac{1}{3}$ of a right-angle?
5. 10 holes are equally spaced on the circumference of a circle. Find the angle between 2 adjacent holes.
6. Add together (a) 19° and 36° (b) 17° 23′ and 16° 11′

Fig. 1.68

11. Convert 36° 42′ to degrees and decimals of a degree.
12. Convert the following to decimals of a degree correct to two decimal places: (a) 27° 18′ 30″ (b) 49° 27′ 25″ (c) 64° 49′ 50″ (d) 11° 17′ 8″.
13. Convert 18·7° to degrees and minutes.
14. Convert the following to degrees, minutes and seconds: (a) 75·8° (b) 3·56° (c) 37·825°.

1.14   Triangles

Types of triangles

(1) *Acute-angled* in which all the angles are less than 90°
(Fig. 1.69a).
(2) *Obtuse-angled* in which one angle is greater than 90°
(Fig. 1.69b).
(3) *Right-angled* in which one angle equals 90°
(Fig. 1.69c).
(4) *Equilateral* which has all three sides equal
(Fig. 1.69d).
(5) *Isosceles* which has two sides equal
(Fig. 1.69e)

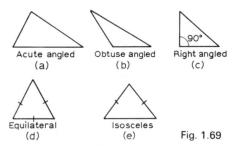

Acute angled   Obtuse angled   Right angled
(a)   (b)   (c)

Equilateral   Isosceles
(d)   (e)   Fig. 1.69

Angle properties of triangles

(1) The sum of the three angles of any triangle is 180°
(Fig. 1.70a).
(2) In an equilateral triangle the three angles are all equal
and hence each is 60° (Fig. 1.70b).
(3) In an isosceles triangle the angles opposite the equal
sides are themselves equal (Fig. 1.70c).
(4) The smallest angle of a triangle lies opposite the
shortest side. The largest angle lies opposite the longest
side (Fig. 1.70d). Angle A+angle B+angle C = 180°.

The three angles of an equilateral triangle are each equal
to 60°.

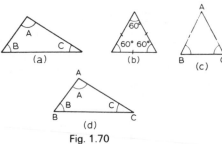

(a)

(b)

(c)

(d)

Fig. 1.70

In the isosceles triangle shown, the side AB equals
the side AC. The angles ABC and ACB which lie opposite
these equal sides are themselves equal.

The angle C is the smallest angle. The side AB lies
opposite the angle C and hence it is the shortest side.
The angle A is the largest angle and the side BC lies
opposite it. Hence BC is the longest side.

Properties of the isosceles triangle. A perpendicular
dropped from the apex to the unequal side:
(1) bisects the unequal side. Thus in Fig. 1.71 BD = CD.
(2) bisects the apex angle. Thus in Fig. 1.71 ∠BAD =
∠CAD.

Fig. 1.71

Example. In a right-angled triangle the two sides forming
the right-angle are equal in length. What are the remaining
angles of the triangle?
Since the sum of the angles of a triangle is 180°, in
Fig. 1.72
    angle A+angle B = 180°−90°
or angle A+angle B = 90°
Since the triangle is isosceles
    angle A = angle B
∴ Both angle A and angle B = 45°

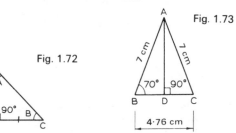

Fig. 1.72

Fig. 1.73

Example. For the triangle shown in Fig. 1.73 find
(a) the angle BCA
(b) the angle BAC
(c) the angle BAD
(d) the length CD

(a) The triangle is isosceles since AB = AC

∴ ∠ABC = ∠BCA

∴ ∠BCA = <u>70°</u>

(b) Since the sum of the angles of a triangle is 180°,

∠BAC = 180°−70°−70° = <u>40°</u>

(c) Because AD is perpendicular to the base ∠BAC is bisected.

∴ ∠BAD = $\frac{40°}{2}$ = <u>20°</u>

(d) Because the base is bisected by the perpendicular

CD = $\frac{4·76}{2}$ = <u>2·38 cm</u>

Properties of the right-angled triangle

(1) In a right-angled triangle the side which lies opposite the right-angle is called the *hypotenuse*. This side is the longest side of a right-angled triangle.

(2) The square on the hypotenuse is equal to the sum of the squares on the other two sides. This statement is known as the theorem of Pythagoras. As shown in Fig. 1.74:

The square on the hypotenuse = AC²

The square on the side AB   = AB²

The square on the side BC   = BC²

∴ AC² = AB²+BC²

We can use this theorem for many calculations which involve right-angled triangles.

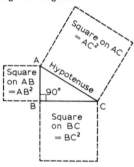

Fig. 1.74

Example. In a right-angled triangle the two sides forming the right-angle are 6 cm and 8 cm long respectively. Find the length of the hypotenuse. From Fig. 1.75,

AC² = AB²+BC²

∴ AC² = 6²+8² = 36+64 = 100

∴ AC = $\sqrt{100}$ = <u>10 cm</u>

Fig. 1.75

Example. A rectangle is 9 cm long and 5 cm wide. What is the length of the diagonal of the rectangle?

From Fig. 1.76 it will be seen that we have to find the length AC, which is the hypotenuse of the triangle ABC.

AC² = AB²+BC²

AC² = 5²+9² = 25+81 = 106

∴ AC = $\sqrt{106}$ = <u>10·3 cm</u>

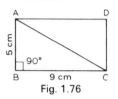

Fig. 1.76

(The square root in this case is found by the method shown in Chapter 1.5.)

Example. A square is to be milled on a round bar 40 mm diameter. Find the side of the largest square that can be obtained.

Fig. 1.77 shows the largest square. It will be seen that the diagonal of the square equals the diameter of the bar. In the triangle ABC,

AB²+BC² = AC²

AB²+BC² = 40² = 1600

Since ABCD is a square,

AB = BC and AB² = BC²

∴   AB²+AB² = 1600

or 2×AB² = 1600

AB² = 800

∴   AB = $\sqrt{800}$ = <u>28·3 mm</u>

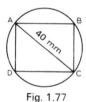

Fig. 1.77

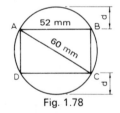

Fig. 1.78

Example. The rectangular shape shown in the diagram has to be milled on a round bar which is 60 mm diameter. Find *d* the depth of cut which has to be taken.

From Fig. 1.78 it can be seen that the diagonal of the rectangle equals the diameter of the bar. We first find BC before we can find *d*.

In the triangle ABC,

AB²+BC² = AC²

or BC² = AC²−AB² = 60²−52² =

3600−2704 = 896

BC = $\sqrt{896}$ = 29·93 mm

2d+BC = 60

or 2d = 60−BC = 60−29·93 = 30·07

d = $\frac{30·07}{2}$ = <u>15·035 mm</u>

Exercise 1.48

1. Find the third angle of each of the triangles shown in Fig. 1.79.

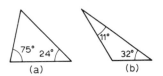

(a)   (b)   (c)

Fig. 1.79

2. Fig. 1.80 shows some isosceles triangles with one angle shown. What are the remaining angles?

(a)   (b)   (c)

Fig. 1.80

3. In the isosceles triangles ABC shown in Fig. 1.81:
   'find (a) the angle BAC
       (b) the angle BAD
       (c) the length of BD

Fig. 1.81   Fig. 1.82

4. In the isosceles triangle shown in Fig. 1.82 the sides AB and AC are equal and ∠BAC = 42°. Find
   (a) ∠ABC   (b) ∠DAC   (c) DC

5. Find the length of the hypotenuse for each of the right-angled triangles shown in Fig. 1.83.

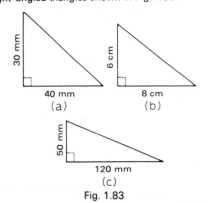

Fig. 1.83

6. Find the length of the hypotenuse for each of the right-angled triangles shown in Fig. 1.84.

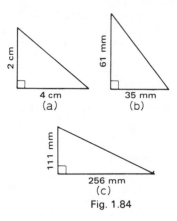

(a)   (b)

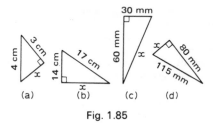

256 mm
(c)

Fig. 1.84

7. Find the side marked x for each of the right-angled triangles shown in Fig. 1.85.

(a)   (b)   (c)   (d)

Fig. 1.85

8. Find the diagonal of a square of side 30 mm.
9. Find the diagonal of a rectangle which is 8 cm long and 5 cm wide.
10. A square is to be milled on a round bar 50 mm diameter. What is the side of the largest square than can be obtained?
11. Find the distance across the corners of the square-headed nut shown in Fig. 1.86.

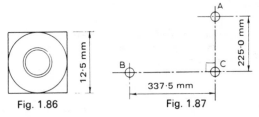

Fig. 1.86   Fig. 1.87

12. An isosceles triangle has a base of 8 cm and a height of 4·5 cm. Find the lengths of the equal sides.
13. A rectangular piece is to be machined from a round bar. If it is to be 20 mm × 40 mm what is the diameter of the smallest bar that can be used?
14. Fig. 1.87 shows three holes which are to be drilled in a plate. The distance between holes A and B is needed for checking purposes. Calculate this distance.

Part 2   Craft science

2.1   Forces

Effect of a force. We cannot see a force but we can see the effect that a force has. We use a force to push the tailstock of a lathe along the bed. We see the effect of the force by noting that the tailstock moves. Again, if we support an object by means of a rope, the rope must supply a force to keep the object in position. We know that this is so, because if the object is not supported it will fall to the ground.

Units of force. Forces are measured in newtons (abbreviation: N), kilonewtons (kN) or meganewtons (MN).
1 000 newtons = 1 kilonewton
1 000 000 newtons = 1 meganewton

Mass. Mass is the amount of matter which is contained in an object. The mass of an object may be found by using an ordinary set of balancing scales. We place the object on one scale and known masses in the other. When the scales balance the mass of the object is equal to the known masses. Masses are measured in kilograms (kg).

Weight. A mass which is dropped will fall to the ground because of the force which the earth exerts on it. This force is called the *force of gravity*. If you hold a book in your hand you prevent the book falling to the ground by exerting a force equal and opposite to the force of gravity.

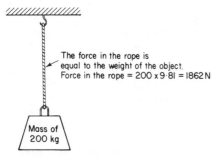

The force in the rope is equal to the weight of the object.
Force in the rope = 200 x 9·81 = 1862 N

Mass of 200 kg

Fig. 2.1

The weight of an object is equal to the *force* which the earth exerts on it. Since weight is a force it is measured in newtons. There is a connection between the mass of an object and its weight which is
weight (in newtons) = mass (in kilograms) × 9·81

Example. A body has a mass of 100 kg. What is its weight?
weight = 100×9·81 = <u>981 newtons</u>

In Fig. 2.1 a mass of 200 kg is supported on the end of a rope. The rope prevents the mass from falling to the ground and hence the force in the rope is equal to the force which the earth exerts on it. That is, the force in the rope is equal to the weight of the mass. The force in the rope is 200×9·81 = 1862 newtons.

When we lift an object—against gravity—we must exert a force equal to the weight of the object. Thus to lift an object having a mass of 20 kg we must apply a force of 9·81×20 = 196·2 N.

Types of forces

1. *Tensile forces.* When we pull a piece of elastic it stretches. Whenever a force is applied which tends to increase the length of an object we say that a tensile force has been applied. When an object is supported by a rope as in Fig. 2.1, the rope tends to increase in length and hence it contains a tensile force. The rope itself is said to be in *tension*.

2. *Compressive forces.* When we push on a spring it shortens and we say that we have applied a compressive force to it. Whenever a force is applied which tends to shorten an object we say that a compressive force has been applied. The legs of a surface table support the weight of the table, work and equipment. In doing so they contain compressive forces and are said to be in *compression*.

3. *Shearing forces.* When a force is applied which causes one layer of material to tend to slide over another layer, we say that a shearing force has been applied. Bolts, rivets and pins are usually subjected to shearing forces.

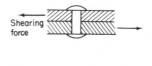

Shearing force

Rivet has sheared because the shearing forces are greater than the rivet can withstand

Fig. 2.2

Fig. 2.22 shows a rivet which has shearing forces applied to it. If these forces are large enough the rivet will shear, because one part of the rivet will slide over the other.

Effects of forces on engineering materials

1. *Tension.* When metals have a tensile force applied to them they stretch like any other material. The amount of stretch depends upon the area of the cross-section of the metal, the type of metal and the size of the force applied. If the force applied is not too large the metal will return to its original size when the force is removed. If too large a force is applied the metal will become permanently distorted. Of course if the force is large enough the metal breaks. Generally speaking, we try to make articles so that they neither permanently distort or break.

2. *Compression.* We have to be very careful when dealing with compressive forces. If a very short bar is placed in compression it will shorten under the action of the forces. A long thin bar, however, will break because of buckling rather than compression, as shown in Fig. 2.3. This buckling effect is the main cause of small drills breaking, because the feed force is too great.

Fig. 2.3   A long thin bar breaks because of buckling
caused by a compressive force

3. *Bending.* The beams shown in Fig. 2.4 bend under the action of forces applied to them. As shown, one edge of the beam is in tension, whilst the other is in compression. The amount of bending (or deflection) depends upon the size of the force and the dimensions of the beam. A large force will cause more deflection than a small one, whilst a deep beam will bend less than a shallow one (Fig. 2.5). The beam in Fig. 2.4 (a) is similar to a bar being turned between centres whilst in Fig. 2.4 (b) the bar is being held in a chuck. The use of steadies prevents the bar bending whilst the cut is being taken.

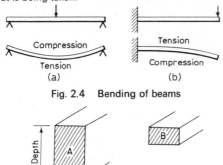

Fig. 2.4   Bending of beams

Fig. 2.5   Beam A will bend less than beam B because
it has a greater depth

Notching a beam (Fig. 2.6) will make it easier to bend, because at the notch the depth of the beam has been reduced.

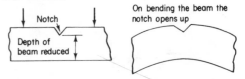

Fig. 2.6   Effect of notching a beam

Beams are often made from channel or **I**-sections. One flange of the section then takes all the tension whilst the other flange takes all the compression (Fig. 2.7). The web between the flanges takes only a little of the bending forces. Drilling holes in the flanges of such beams can seriously weaken them. If holes have to be drilled they should be drilled in the web of the beam since this causes far less weakening of the beam.

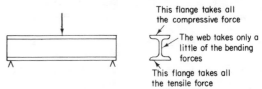

Fig. 2.7   The way in which an **I**-section beam carries
the bending forces

Reinforced concrete beams have steel reinforcing placed in them. Concrete is very strong in compression and weak in tension whilst the steel is very strong in tension. Thus when a reinforced concrete beam bends the concrete takes the bulk of the compressive loading whilst the steel takes most of the tensile loading. It is bad practice to drill holes in concrete beams because some of the steel reinforcing may be cut by the drill thus seriously weakening the beam.

4. *Twisting.* We can use a force to cause rotation. The forces (exerted by the hands) on a tap wrench cause the tap to rotate and cut the thread. A spanner causes a nut to rotate when a force is applied to its end. If, in tightening the nut, we apply too great a force at the end of the spanner we cause the thread on the bolt to twist off. When this happens the thread shears away from the main body of the metal. The key holding a milling cutter on the arbor will shear off if too great a cutting force is applied to the cutter teeth. Shafts, when starting to rotate, twist and if they are not strong enough they will break because of the shearing action which they encounter.

Stress and strain. When a bar of metal is in tension the size of the force it can withstand depends upon the material from which the bar is made and the area of the cross-section of the bar. A bar with a large cross-sectional area can withstand a much larger force than one with a small cross-sectional area. In order to compare the strengths of the two bars we calculate the force per unit area and so comparison is then made easy. The force per unit area is called the *stress*. Thus

$$\text{stress} = \frac{\text{force carried by the bar}}{\text{cross-sectional area}}$$

Stress is measured in newtons per square metre (N/m²) or meganewtons per square metre (MN/m²).

Example. A short bar in compression has a cross-sectional area of 20 square millimetres and it carries a force of 10 000 newtons. Find the stress in the bar.

$$\text{Stress} = \frac{\text{force carried by the bar}}{\text{cross-sectional area}}$$

Cross-sectional area = 20 mm² = $\frac{20}{1\,000\,000}$ m²

$$\text{Stress} = 10\,000 \div \frac{20}{1\,000\,000}$$

$$= \frac{10\,000}{20} \times 1\,000\,000$$

$$= 500 \times 1\,000\,000 \text{ N/m}^2$$

$$= 500 \text{ MN/m}^2$$

In order to compare the strengths of engineering materials the maximum stress that a material can withstand before breaking is used. This maximum stress is usually called the *ultimate stress*. Some values for the ultimate stresses of some common metals are given below.

ULTIMATE STRESSES FOR METALS IN TENSION

Metal	Ultimate stress (MN/m²)
Cast iron	110
Mild steel	400
Aluminium	80
Brass	410
Copper	200

From the table it can be seen that aluminium is a comparatively weak metal whilst mild steel is a much stronger metal.

We have already seen that when a bar of metal has a tensile force applied to it the bar stretches. When a compressive force is applied to it the bar shortens in length. Whenever a force is applied to an object it undergoes some distortion. The amount of distortion is measured by the *strain*. For a material which is in tension or compression the strain is given by the formula:

$$\text{Strain} = \frac{\text{alteration in length}}{\text{original length}}$$

Example. A bar of metal 2500 mm long has a tensile force applied to it. Under the action of the force its length increases by 0·75 mm. What is the strain?

$$\text{Strain} = \frac{\text{alteration in length}}{\text{original length}}$$

$$= \frac{0·75}{2500}$$

$$= 0·0003$$

Notice that strain has no units—it is just a number.

Force as a vector quantity. A vector is a quantity which possesses both size and direction. A force which acts on a body in a particular direction is a vector quantity. Fig. 2.8 shows a force acting on a lathe tool. The force has a size (250 N) and a particular direction (indicated by the angle of 15° and the arrow-head). Again, the weight of an object is a vector quantity, since it is the force exerted by the earth on the object and it acts towards the centre of the earth. If the object is to be lifted upwards a force must be applied vertically upwards to overcome the gravitational force exerted by the earth.

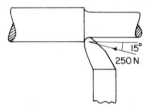

Fig. 2.8   The force acting on a lathe tool possesses both size and direction and hence it is a vector quantity

Frequently an object may be acted on by a number of forces. The object in Fig. 2.9 has two forces acting on it. If the object is free to move, it will move in the direction shown in the diagram and it will move as if only one resultant force were being applied.

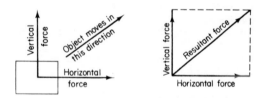

Fig. 2.9   An object with two forces acting on it moves in the direction of the resultant force

The resultant force in Fig. 2.9 is greater than either the horizontal force or the vertical force. In order to find the magnitude of the resultant force we need to be able to represent forces by a drawing.

Representation of forces. To represent a force we need to know the following information about it:
(1) its magnitude (or size);
(2) its direction;
(3) its point of application;
(4) its sense (i.e. does it pull away from the point of application or does it push towards it?). The method used is shown in the following example.

Example. Represent graphically the force shown in Fig. 2.10.

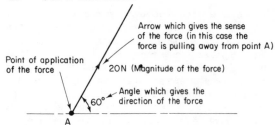

Arrow which gives the sense
of the force (in this case the
force is pulling away from point A)

Point of application
of the force

20N (Magnitude of the force)

Angle which gives the
direction of the force

60°

A

Fig. 2.10   To represent a force we need to know its
magnitude, direction, point of application
and sense

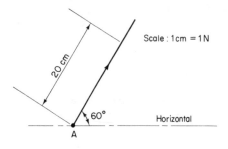

Scale : 1cm = 1N

20 cm

60°

A   Horizontal

Fig. 2.11   The force of Fig. 2.10 represented graphically

The procedure is as follows (Fig. 2.11):
(1) Faintly draw a horizontal line and on it mark the point A which is the point of application of the force.
(2) From A draw a faint line inclined at 60° to the horizontal.
(3) Choose a suitable scale to represent the magnitude of the force. In Fig. 2.11 a scale of 1 cm = 1 N has been chosen. Hence 20 N will be represented by a length of 20 cm. From A mark off a length of 20 cm along the inclined line.
(4) Place an arrow on the line to indicate the sense of the force.
We have now represented the force graphically.

Example. A force of 15 N pushes towards a point X and it acts in a direction which is 45° to the vertical. Represent this force graphically. The method is shown in Fig. 2.12.

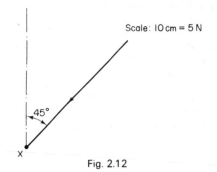

Scale: 10 cm = 5 N

45°

X

Fig. 2.12

(1) The magnitude is represented by a length of 30 cm.
(2) The arrow represents the sense.
(3) The angle at which the line is inclined represents the direction.

Resultant force. In Fig. 2.13, two forces, one of 5 N and the other of 6 N act on a body. Both the forces have the same point of application. A single force of 5 N+6 N = 11 N acting to the right and having the same point of application would have precisely the same effect as the two forces of 5 N and 6 N. This single force of 11 N is called the *resultant force*.

The *resultant* of a system of forces is the single force which produces the same effect as the system of forces.

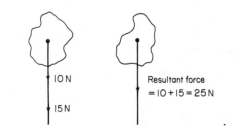

5N   6N

Resultant force

=11N

Fig. 2.13   A single force of 11 N has precisely the same effect as the two forces of 5 N and 6 N

Other examples are shown in Figs 2.14 and 2.15.

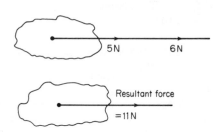

10N

15N

Resultant force
= 10+15 = 25 N

Fig. 2.14

8N   10 N

Resultant force
= 10−8 = 2N

Fig. 2.15

The parallelogram of forces. So far we have considered only forces which act in the same straight line. Frequently we need to find the resultant of forces which have different lines of action (Fig. 2.16). The diagram shows forces of 15 N and 8 N both applied to the point O. If the point O is allowed to move, it will do so in a direction which lies between OA and OB, that is, in the direction OX. Hence a single force acting in the direction OX would have precisely the same effect on the point O as the two forces shown. Of course, the force will have to be of the correct size. This force will then be the resultant force. The parallelogram of forces is a method whereby the resultant force may be found.

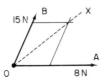

Fig. 2.16   A single force of the correct size acting in the direction OX will have the same effect on the point O as the two forces of 15 N and 8 N

Example. Two forces, one of 10 N and the other of 20 N each pull away from a point P. The two forces are inclined at 45° to each other. Find their resultant force.

(1) Choose a suitable scale and represent the two forces graphically.
(2) Construct a parallelogram with the lines representing the forces as adjacent sides.
(3) Draw the diagonal from P. This diagonal represents the resultant force in magnitude and direction (the diagonal which represents the resultant is always drawn from the point of application of the two forces). By measuring, the resultant force is found to be 28 N acting at 15° to the 20N force.

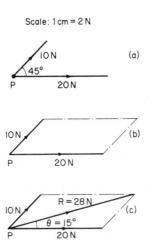

Fig. 2.17   The parallelogram of forces

Example. Find the resultant of the two forces shown in Fig. 2.18.

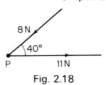

Fig. 2.18

We see from Fig. 2.18 that the 8 N force pushes towards P whilst the 11 N force pulls away from P. Before we can draw the parallelogram of forces *both* arrow-heads must either point towards P or they must *both* point away from P. Consider the force F shown in Fig. 2.19. It does not matter if F pushes on the body or if it pulls on the body—the effect on the body will be the same in either case.

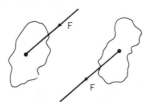

Fig. 2.19   The effect on the body is the same whether F pushes on it or whether it pulls on it

Hence we may alter Fig. 2.18 in the ways shown in Fig. 2.20 without altering the effect the forces have on the point P. We can use either diagram to construct the parallelogram as shown in Fig. 2.21. Both constructions produce the same answer, R = 7·25 N and θ = 48°. Notice that in Fig. 2.21 all the arrow-heads either point towards P or point away from P.

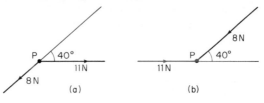

Fig. 2.20

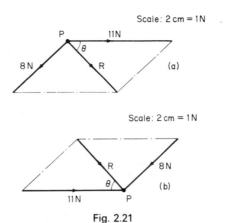

Fig. 2.21

Sometimes we have to split a force up into component parts. The method is shown in the following example.

Example. A casting weighing 1000 N is supported by two ropes fastened to eye-bolts 3 metres apart. The ropes are each 2·5 metres long. Find the forces in each of the ropes.

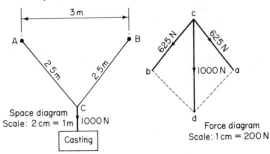

Space diagram
Scale: 2 cm = 1m

Force diagram
Scale: 1 cm = 200 N

Fig. 2.22

The force of 1000 N may be considered to be the resultant of the forces in each of the ropes. The method of solving the problem is as follows:

(1) Draw the space diagram to scale.

(2) Choose a suitable scale for the force diagram and draw a vertical line cd to represent the weight of the casting. From c draw ca parallel to AC and cb parallel to BC and complete the parallelogram.

(3) Scale ac and bc and thus find the forces in the ropes. These are found to be 625 N.

When supporting articles by means of ropes and slings it is important to realize that the greater the angle between the ropes the greater are the forces in the ropes (Fig. 2.23).

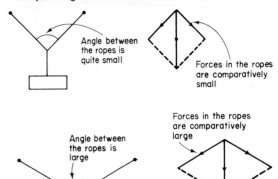

Fig. 2.23   The greater the angle between the ropes the greater the forces in the ropes

Exercise 2.1

1. How many kilograms is 1 megagramme?

2. How many newtons is 1 kilonewton?

3. A body has a mass of 500 kg. What is its weight?

4. A small casting has a mass of 90 kg. What is the weight of the casting?

5. If a shaft having a mass of 2000 kg is supported by a single rope, what is the force (or load) in the rope?

6. What kind of force exists in the rope of question 5?

7. A floor is supported by columns. What kind of force is there in the columns?

8. When a compressive force is applied to a long slender object what happens to it?

9. What kind of force is there in the bolt shown in Fig. 2.24?

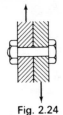

Fig. 2.24

10. Fig. 2.25 shows a beam with a force acting on it. Which edge of the beam is in tension?

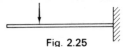

Fig. 2.25

11. What is the effect of notching a beam?

12. Why should holes be drilled in the web of an **I**-beam but not in the flanges?

13. Why must holes not be drilled in reinforced concrete beams?

14. What is the effect of applying too large a force to the end of a spanner when tightening up a nut?

15. A bar in tension has a cross-sectional area of 25 square millimetres. If the force applied is 1000 newtons what is the stress in the bar?

16. A short bar carries a compressive force of 5000 N. The cross-sectional area of the bar is 40 mm². What is the stress in the bar?

17. What is meant by ultimate tensile stress?

18. What is strain?

19. A bar of metal 1500 mm long has a tensile force applied to it. Under the action of this force the length of the bar increases by 1·5 mm. What is the strain?

20. A wire 3 metres long and having a cross-sectional area of 0·06 mm² has a force of 30 N applied to it. Under the action of this force it stretches 0·10 mm. Find the stress in the wire and also the strain.

21. Why is force a vector quantity?

22. Represent graphically the following forces:
   (a) 40 N acting at 30° to the horizontal pulling away from a point O.
   (b) 80 N acting at 60° to the vertical pushing towards a point O.
   (c) 2000 N acting at 45° to the horizontal pushing towards a point O.

23. Find the resultant of the systems of forces shown in Fig. 2.26.

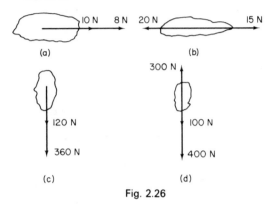

Fig. 2.26

24. Find the resultants of the forces shown in Fig. 2.27.

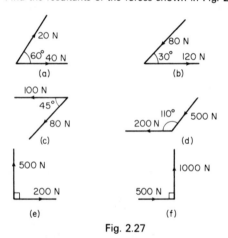

Fig. 2.27

25. Fig. 2.28 shows the forces at the cutting point of a lathe tool. Find the resultant force on the tool.

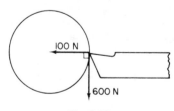

Fig. 2.28

26. Fig. 2.29 shows a casting supported by two ropes. If the maximum force in each of the ropes must not exceed 500 N, what is the greatest weight of casting that can be supported? What mass can be supported?

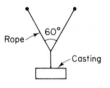

Fig. 2.29

27. Fig. 2.30 shows an object being supported on two ropes. What is the force in each rope?

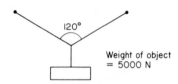

Fig. 2.30

28. Which of the arrangements in Fig. 2.31 puts the sling in the greatest tension?

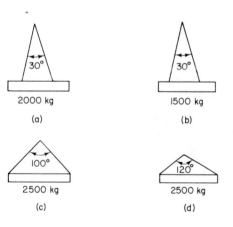

Fig. 2.31

2.2 Moments

The moment of a force. Fig. 2.32 shows a force being applied to a door. The door rotates about its hinge so that the force has had a *turning effect* on the door. Another way of saying this is to say that the door has a *moment* applied to it. The point about which the object turns is often called the *fulcrum*.

Door

Hinge

Fig. 2.32

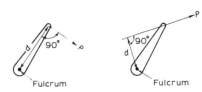

Fulcrum

Fulcrum

Fig. 2.33

The moment of a force about a fulcrum is measured by the product of the force and its distance from the fulcrum. The distance from the fulcrum must be measured perpendicular to the line of action of the force as shown in Fig. 2.33. In both cases the moment is $P \times d$.

Figs 2.34 (a) and (b) show a lever hinged at a bracket. In each case the 100 newton force causes the beam to rotate at (a) in a clockwise direction and at (b) in an anti-clockwise direction. Moments therefore may be either clockwise or anti-clockwise.

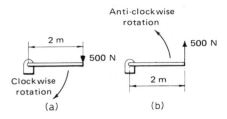

Fig. 2.34

(a) Clockwise moment = 500×2 = 1000 newton metres
(b) Anti-clockwise moment = 500×2 = 1000 newton metres

The units of moment depend on the units of force and the units of distance as follows:
Force in newtons—distance in millimetres—moment in newton millimetres (N mm).
Force in newtons—distance in metres—moment in newton metres (N m).

Example. A tommy bar is used to operate a box spanner (see Fig. 2.35). What turning moment do the forces produce?

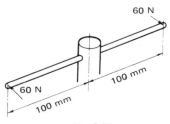

Fig. 2.35

Turning moment = $(60 \times 100) + (60 \times 100)$
= 6000+6000
= 12 000 N mm (clockwise)

Principle of moments. Referring to Fig. 2.36:
If the turning moment produced by P (i.e. Py) is greater that that produced by Q (i.e. Qx) then the lever will rotate in a clockwise direction. If however the moment produced by Q is greater than that produced by P then the lever will rotate in an anti-clockwise direction. When the moments produced by P and Q are the same the lever will balance. Now Py is a clockwise moment and Qx is an anti-clockwise moment so the lever balances when the clockwise moments equal the anti-clockwise moments (i.e. when $Px = Qy$). *For balance (equilibrium) the sum of the clockwise moments must equal the sum of the anti-clockwise moments.* This statement is known as the Principle of Moments.

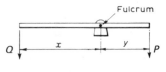

Fig. 2.36

Examples

1. Find the force *F* required to balance the lever shown in Fig. 2.37.

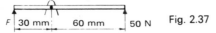

Fig. 2.37

Clockwise moment   = $F \times 30 = 30 F$
Anti-clockwise moment = $50 \times 60 = 3000$ N mm
For balance, the anti-clockwise moment must equal the clockwise moment.
$\therefore 30 F = 3000$
     $F = \underline{100 \text{ newtons}}$

2. Fig. 2.38 shows a pair of pincers. Find the gripping forces on the work.

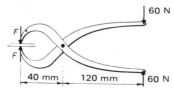

Fig. 2.38

Considering one of the levers we have
Anti-clockwise moment = $60 \times 120 = 7200$ N mm
Clockwise moment   = $F \times 40 = 40 F$
     $40 F = 7200$
       $F = \underline{180 \text{ newtons}}$

Reaction at the fulcrum. The lever shown in Fig. 2.39 will not rotate since the anti-clockwise moment ($4 \times 1000 = 4000$ N m) equals the clockwise moment ($2 \times 2000 = 4000$ N m). But what stops the lever falling to the ground? Of course the fulcrum pin and bracket prevent this. Therefore the fulcrum pin must support the total load ($1000 + 2000 = 3000$ N) on the lever. This load on the pin of 3000 N is known as the reaction at the pin. We can always find the reaction at the fulcrum of a lever if we remember that *the total upward forces equal the total downward forces*.

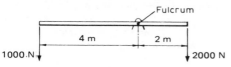

Fig. 2.39

Example. The lever shown in Fig. 2.40 is used for testing springs. Calculate the force on the spring and the force on the pivot when a load of 200 N is applied as shown.

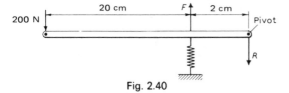

Fig. 2.40

Let *F* = force on the spring
Anti-clockwise moment about the pivot
     $= 200 \times 22 = 4400$ N cm
Clockwise moment about the pivot
     $= F \times 2 = 2 F$ newton centimetres
$\therefore$ $2 F = 4400$
     $F = \underline{2200 \text{ newtons}}$
Let *R* = force on the pivot
Total downward forces = $(200 + R)$ newtons
Total upward force = $F = 2200$ newtons
     $\therefore 200 + R = 2200$
         $R = 2200 - 200$
         $= \underline{2000 \text{ newtons}}$

Example. Fig. 2.41 shows a job clamped to the table of a machine. If the clamping bolt provides a force of 1500 N, find the force holding the work in position.

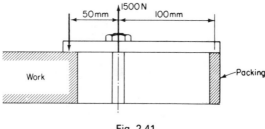

Fig. 2.41

We can solve this problem by treating the packing as the fulcrum and then we get a lever like that shown in Fig. 2.42.

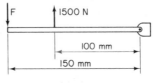

Fig. 2.42

Taking moments about the fulcrum
Anti-clockwise moment = $F \times 150 = 150 F$
Clockwise moment   = $150 \times 100 = 150\,000$
     $\therefore 150 F = 150\,000$
         $F = \underline{1000 \text{ newtons}}$

Notice that the force holding the work in position (*F* in Fig. 2.42) is less than the force supplied by the clamping bolt. It pays, when clamping, to place the clamping bolt as near the work as possible in order to obtain a large force to hold the work in position.

Centre of gravity. The centre of gravity of a body is the point at which all the mass of the body may be assumed to be concentrated. For thin plates the centre of gravity is always the geometrical centre of the plate. Thus,

the c.g. of a circle is at the centre of the circle,
the c.g. of a rectangle is at the intersection of the
diagonals.

For a solid bar of metal of constant cross-section, the
c.g. lies exactly half-way along the axis of the bar.

Balance and unbalance. A body will only be balanced
if it is supported at its c.g. If it is supported anywhere else
it will be in a state of unbalance.

Example. A uniform bar of metal 4 metres long and
weighing 180 newtons is supported at 1 metre from
one end. Find the force needed to balance the bar if this
force is applied at 0·5 metre from the same end as shown
in Fig. 2.43 (a).

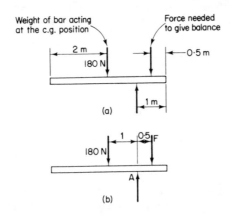

(a)

(b)

Fig. 2.43

The problem reduces to finding the force *F* in Fig.
2.43 (b)

Taking moments about the point A

$$0·5\, F = 180$$

$$F = \frac{180}{0·5} = 360$$

The force required to balance the bar is <u>360 N</u>.

Exercise 2.2

1. In Fig. 2.44 a crowbar is being used to lift a heavy
   object. Find the force *P* which the crowbar applies
   to the object.
2. Find the force *P* required at the ends of the handles
   of the shears shown in Fig. 2.45.

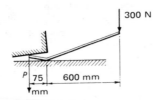

Fig. 2.44

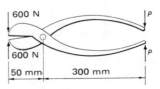

Fig. 2.45

3. Fig. 2.46 shows a simple lever. Find the effort *E*
   needed to balance the load of 80 N. What is the force
   on the pin?

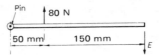

Fig. 2.46

4. Fig. 2.47 shows a bench guillotine. Find the effort
   E needed to supply the shearing force of 1000 N
   which is applied to the work.

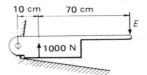

Fig. 2.47

5. A milling cutter is 100 mm diameter and the cutting
   force at the teeth is 2000 N. Find the turning moment
   applied to the spindle.
6. In using a tap wrench the hands are placed 20 cm
   apart. If each hand exerts a force of 40 N what turning
   moment is applied to the taps?
7. For a turning operation, the work is driven by a faceplate pin set at 15 cm from the centre of the work. If
   the work is 75 mm diameter and the cutting force is
   800 N, what is the force on the driving pin?
8. The turning moment needed to drill a 25 mm diameter
   hole was found to be 60 N m. What force is acting at
   each of the cutting edges of the drill?
9. Fig. 2.48 shows a plate clamp
   (a) find the force *P*
   (b) find the tension on the bolt.
10. Complete the missing information in the table
    opposite (refer to Fig. 2.49).

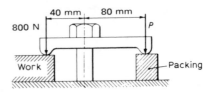

Fig. 2.48

P newton	Q newton	x cm	y cm	R newton
8	?	5	10	?
?	4	2	8	?
10	4	3	?	?
8	6	4	?	?
7	10·5	?	6	?

11. A turning moment of 40 newton metres is to be applied to a nut. Find the force needed at the end of a spanner 20 cm long.
12. A bar of metal 5 m long and weighing 200 N is supported at 2 m from one end. If a force of 500 N is used to balance the bar, where must it be positioned?

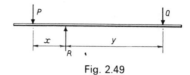

Fig. 2.49

2.3 Temperature

Effects of heat. Forging, soldering, brazing and tempering operations all make use of heat. The application of heat to various materials can effect them in a number of ways:

Change in dimension. Liquids, gases and metals expand when heated and contract when cooled.

Change of state. When ice is heated it melts and becomes water; with further heating it boils and becomes steam. The reverse sequence occurs when steam is cooled; the steam condenses to water which can be frozen to produce ice. During soldering the solder is melted by heating it and it solidifies on cooling.

Change of composition. If sugar is heated it first turns into a brown caramel and then into black carbon. The reverse process is not possible and we cannot obtain sugar by cooling down the carbon.

Change of colour. When a piece of steel is heated in a forge its colour changes. The colour varies from a dark red to cherry red to yellow and then to white. These various colours give us an idea how hot the metal is. We know, from experience, that a white-hot piece of metal is much hotter than a piece which is red hot.

Electrical effect. This is, perhaps the least familiar effect of all. If two wires of dissimilar metals, such as copper and iron, are joined together and heated at the joint a small electrical current is produced. This current can be measured on a sensitive instrument called a *galvanometer.* The junction of the two metals is called a *thermocouple.*

Temperature. This is the degree of hotness or coldness of a body. A body having a high temperature is hot whilst one having a low temperature is cold.

Measuring temperature. Temperature can be measured by observing its effects. Change of dimension, change of state, change of colour and the electrical effect are all used to measure temperature.

Thermometers. A thermometer measures temperature by means of the expansion or contraction of the liquid in the tube. Alcohol is the liquid used for measuring low temperatures, whilst mercury is used for measuring high temperatures.

Temperature scale. When marking the scale of the thermometer it is necessary first to mark two 'fixed points', representing the melting point of ice and the boiling point of water. The positions of the 'fixed points' are found by first placing the thermometer in melting ice and marking on the tube the lowest point reached by the liquid. The thermometer is then placed in steam produced by boiling water at atmospheric pressure, and the highest point reached by the liquid is marked on the tube. The scale is then completed between the two fixed points, using any number of equal divisions.

Temperatures are measured on the Celsius scale. The boiling-point of water is marked at 100° and the freezing point at 0°. There are then 100 divisions between the two 'fixed points'. Each division is 1 degree Celsius written as 1 °C (Fig. 2.50).

Measuring high temperatures. Many of the operations which take place in the workshop need very high temperatures—much higher than can be measured with a thermometer. An instrument called a *pyrometer* is used for measuring these high temperatures. In its operation

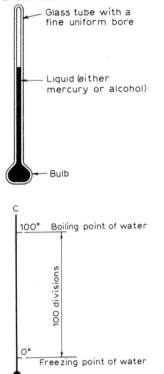

Fig. 2.50   The Celsius thermometer

it makes use of the electrical effect produced by heat (using a thermocouple) and it is capable of measuring temperatures up to about 1400 °C. Fig. 2.51 shows how the instrument works. (Temperature can be very accurately measured with this instrument.)

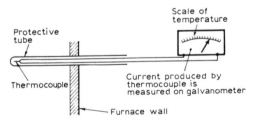

Fig. 2.51   A thermocouple pyrometer.

Another type of pyrometer is the *optical pyrometer*. The way in which this instrument works is shown in Fig. 2.52. Electrical current is used to heat a wire. As the wire heats up it changes colour (see table of heat colours below) and the colour of the wire is compared with the heat colour of the hot body. The current is adjusted until the colour of the glowing wire is the same as the colour of the hot body, which means that the temperatures of the two are the same. This temperature is then read off the scale of the instrument. Optical pyrometers give very accurate temperature measurements.

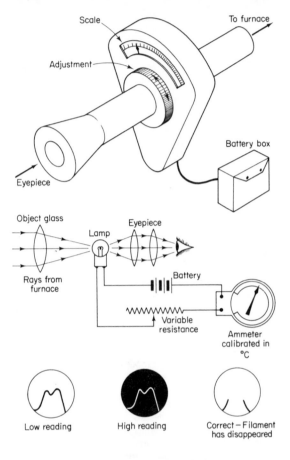

Fig. 2.52   The optical pyrometer

Colour and temperature. The temperature of a hot metal can be judged by its colour. The table below gives the approximate temperature corresponding to the various colours of heated steel.

COLOUR TABLE

Colour	Temp. °C
Dull red	700
Cherry red	900
Orange red	1000
Yellow	1100
White	1300

Heat-sensitive crayons and paints. These may be used for estimating temperatures. They are useful where the small-scale heat treatment of steel parts is to be carried out. The paints and crayons change in colour or appearance at a stated temperature. A mark is made on the component which is then heated until the change in appearance or colour takes place. Paints are useful for indicating the temperature of a component which has to be heated locally, e.g. for pipe bending.

Fusible cones and pellets. *Seger cones* are triangular pyramids which are made from a mixture of kaolin, feldspar, quartz, magnesia and other substances. The composition is adjusted so that the cones wilt and finally collapse at a fixed temperature. The method of using the cones is shown in Fig. 2.53. Temperatures between 600 °C and 2000 °C may be indicated in this way. Seger cones are not very accurate because the rate of heating and the atmosphere of the furnace affect their wilting temperature.

Indicating pellets which are made from fusible salts are used for measuring temperatures up to 900 °C. They melt sharply at their stated temperature with an error of not more than 1 °C.

Fig. 2.53   Taking the temperature of a furnace by using Seger cones. The temperature is estimated from the condition of the cones after they have been in the furnace for a sufficient length of time for them to be affected by the heat.

Both Seger cones and indicating pellets depend upon a change of state (from solid to liquid) for indicating temperature.

Expansion of metals. When a bar of metal is heated it increases in length. The amount the bar increases in length depends upon the length of the bar, the rise in temperature and the kind of metal from which the bar is made. By doing experiments with bars of different metals to study their expansions, it is found that each metal increases in length by a constant amount for each degree of temperature rise. The amount that a bar of unit length (i.e. 1 cm or 1 m) increases in length for each 1 ° of temperature rise is called the *coefficient of linear expansion*. Some values for this coefficient are shown in the table.

COEFFICIENTS OF LINEAR EXPANSION

Metal	Coefficient per 1 °C
Aluminium	0·000 023 8
Brass	0·000 020 0
Copper	0·000 016 7
Cast iron	0·000 010 6
Steel	0·000 011 2
Wrought iron	0·000 011 2

We can calculate the increase in length of a bar by using the formula,

Increase in length = Original length × Coefficient of expansion × Temperature rise

Examples

1. A steel bar 400 mm long is raised in temperature from 15 °C to 100 °C. Calculate the increase in length.
   From the table the coefficient of linear expansion for steel is 0·000 011 2 per 1 °C. The temperature rise is 100 °C−15 °C = 85 °C.
   ∴ Increase in length = 400×0·000 011 2×85 mm
   = 0·381 mm

2. A steel plug gauge 40·00 mm diameter is raised in temperature from 20 °C to 27 °C. What is the diameter of the plug gauge at 27 °C?
   The coefficient of linear expansion for steel is 0·000 011 2 per 1 °C. The temperature rise is 27 °C−20 °C = 7 °C. Hence,
   Increase in diameter = 40×0·000 011 2×7 mm
   = 0·0031 mm
   Diameter at 27 °C   = 40·00+0·0031 = 40·0031 mm
   Although this is a very small increase in diameter it can be important if the plug gauge is used where the tolerances are very small.

Exercise 2.3

1. Write down three effects of heat.
2. How does a thermometer work?
3. Briefly describe the optical and thermocouple pyrometers stating the heat effect on which they work.
4. What heat effect is utilized with Seger cones?
5. A brass scale is 50 cm long at 15 °C. What is its length when the temperature is 30 °C?
6. The length of an iron casting is 80 cm at 810 °C. What is its length at 30 °C?
7. A steel ring gauge is exactly 50 mm diameter at 15 °C. What is its diameter when the temperature is 25 °C?
8. A steel vernier caliper gives a reading of 182·34 mm at 10 °C. If the instrument gives a true reading at 18 °C, find the correct reading.

2.4 Heat

Transmission of heat. Heat can be transferred from one object to another. It may also be transferred from one part of an object to another part of the same object. There are three ways in which heat can be transferred, by radiation, conduction and convection.

1. Radiation (Fig. 2.54). A person in full view of a fire will get warmer than someone who is shielded from it,

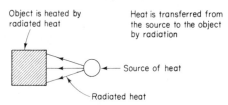

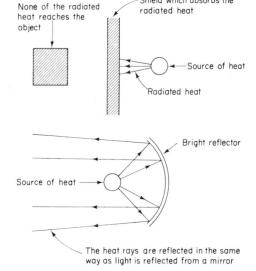

Fig. 2.54   Radiated heat

even although both are sitting at the same distance from the fire. Since neither are touching the fire this kind of heat must be transferred in the same way as light because if the light from the fire is shielded so is the heat. This sort of heat transfer is called *radiation* and heat transferred by this method is called *radiant heat*. Heat from the sun reaches the earth by radiation.

2. Conduction (Fig. 2.55). If we heat one end of a copper bar, the other end of the bar soon becomes hot also. This means the heat from the hot end has been

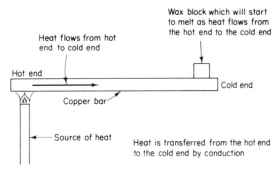

Fig. 2.55   Demonstration of conducted heat

conducted along the bar. Metals are good conductors of heat and we say that their thermal conductivity is good. Some metals conduct heat better than others. Copper is particularly good in this respect and we say that copper has a high thermal conductivity. Other materials such as wood, paper and glass are poor conductors of heat.

3. Convection (Fig. 2.56). If cold air is passed over a heating element it gets warmer. As the air is heated it

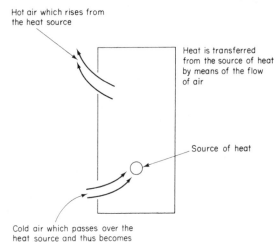

Fig. 2.56   Convected heat

expands and in so doing it becomes lighter. These lighter particles of air rise, thus allowing more cold air to pass over the heating element. Thus a continuous stream of warm air is formed. The heat is transferred from the heating element by means of this flow of air and the process is called *convection*. The flow of air is often called a 'convection current'. Convection currents in liquids cause heat to be transferred to various parts of the liquid.

Heat energy. When a body is heated it is given heat energy. The amount of heat energy an object possesses depends upon:
(1) the mass of the object;
(2) the temperature of the object;
(3) the material from which the object is made.

A small soldering iron (which has a small mass) gives out heat for a shorter time than a large soldering iron (which has a large mass) even although they are both heated to the same temperature. If both soldering irons are raised to a higher temperature then they both give out heat for a longer time.

It takes a much shorter time to heat a quantity of milk which has a mass of 1 kg than it does to heat the same mass of water, both substances reaching the same temperature.

When a body is heated it gains heat energy. When it is cooled it loses heat energy. The units of heat energy are the joule (abbrev: J) or the kilojoule (kJ).

1000 joules = 1 kilojoule

Specific heat. By performing careful experiments it has been found that it needs 4·18 kilojoules of heat energy to raise the temperature of 1 kilogram of water by 1 °C. Similarly it has been found that copper needs only 0·39 kilojoules of heat energy to raise the temperature of 1 kilogram by 1 °C.

The quantity of heat energy needed to raise the temperature of 1 kilogram of the substance by 1 °C is called the *specific heat* of the substance. Thus we say that the specific heat of water is 4·18 kilojoules per kilogram per °C whilst that of copper is 0·39 kilojoules per kilogram per °C. A table of specific heats is given below:

TABLE OF SPECIFIC HEATS
IN KILOJOULES PER KILOGRAM PER °C

Aluminium	0·92	Tin	0·23
Brass	0·39	Zinc	0·39
Copper	0·39	Cast iron	0·54
Iron	0·46	Water	4·18
Lead	0·13		

For any material:
gain in heat energy = mass of material (in kg)
× specific heat (in kJ/kg/°C) × temperature rise (in °C)
loss of heat energy = mass of material (in kg)
× specific heat (in kJ/kg/°C) × temperature drop (in °C)

Examples
1. 2 kg of steel are heated from 20 °C to 800 °C. Find the heat energy gained by the steel.
   Specific heat of steel = 0·46 kJ/kg/°C
   Temperature rise   = 800 °C−20 °C = 780 °C
   Mass of steel   = 2 kg
   Gain in heat energy = 2×0·46×780 = <u>717·6 kJ</u>
2. A copper soldering iron has a mass of 250 g. The temperature of the iron is 200 °C before soldering starts and after the soldering is completed its temperature is 170 °C. How much heat energy has been lost by the soldering iron?
   Specific heat of copper = 0·39 kJ/kg/°C
   Temperature drop   = 200 °C−170 °C = 30 °C
   Mass of copper   = 250 g = 0·25 kg
   Loss of heat energy = 0·25×0·39×30 = <u>2·93 kJ</u>
3. A steel tank contains 500 kg of water. The tank itself has a mass of 80 kg and it is made of steel. If the tank is heated from 15 °C to 65 °C how much heat energy is needed?
   Specific heat of water   = 4·18 kJ/kg/°C
   Temperature rise   = 65 °C−15 °C = 50 °C
   Mass of water   = 500 kg
   Heat gained by the water   = 500×4·18×50
                              = 104 500 kJ
   Specific heat of steel   = 0·46 kJ/kg/°C
   Temperature rise   = 65 °C−15 °C = 50 °C
   Mass of steel   = 80 kg
   Heat gained by steel tank = 80×0·46×50 = 1840 kJ
   Total heat energy needed   = 104 500+1840
                              = <u>106 340 kJ</u>

Sources of heat energy. There are several sources of heat energy but in the workshop the three most important are the flame, the electric arc and the electric resistance.
1. The flame is used for forging processes and for oxy-acetylene welding.
2. The electric arc is used during electric welding.
3. The electric resistance is used, for instance, in an electric fire where the electric current passing through a wire causes the wire to heat up owing to the resistance of the wire to the passage of the electric current.

Each of these will be discussed in greater detail when dealing with forging, welding and the electric current.

Exercise 2.4
1. State three ways in which heat may be transmitted.
2. Name three materials which are good conductors of heat and three which are not.
3. Define (a) the joule (b) the kilojoule.

4. 20 kg of water are heated from 20 °C to 90 °C. What quantity of heat is absorbed by the water? (sp. ht. of water = 4·18 kJ/kg/°C)

5. A steel bar having a mass of 5 kg is heated during heat treatment from 20 °C to 820 °C. How much heat does the steel absorb? (sp. ht. of steel = 0·46 kJ/kg/°C)

6. A copper soldering iron has a mass of 1·5 kg. It is heated to 250 °C. After soldering the temperature falls to 200 °C. How much heat has the soldering iron given up? (sp. ht. of copper = 0·39 kJ/kg/°C)

7. 3900 joules of heat energy are used in heating 50 kg of copper from 20 °C to 220 °C. Calculate the specific heat of copper.

8. Find the quantity of heat lost by a piece of steel having a mass of 20 kg when it is cooled from 1020 °C to 20 °C. (sp. ht. of steel = 0·46 kJ/kg/°C)

9. Name three sources of heat energy giving one workshop use for each heat source.

10. When is convected heat used?

2.5   Friction

The causes of friction. When we attempt to move one surface over another surface as shown in Fig. 2.57 a resistance is set up which prevents (or tends to prevent) movement. This resistance to movement is said to be the result of *friction* between the two surfaces in contact.

Fig. 2.57 (a) shows two machine blocks in contact with each other. Although the two faces in contact appear quite flat to the naked eye, a powerful microscope reveals that there are irregularities in the surfaces as shown at (b). The crests of the irregularities foul each other and the resistance to motion is thus created. The block A will not move over block B until the force P is greater than the frictional resistance.

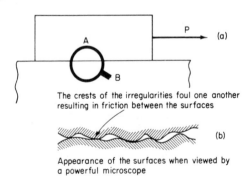

The crests of the irregularities foul one another resulting in friction between the surfaces

Appearance of the surfaces when viewed by a powerful microscope

Fig. 2.57   The cause of friction

The recognizable effects of friction. It will be seen from Fig. 2.57 (b) that actual contact between the faces of the two blocks takes place on only a few very small areas. Even when quite small forces hold the surfaces together very high pressures are created on the areas in contact. (This is rather like a young lady who is comparatively light in weight walking across a floor in stilleto heels. The small area in contact with the floor causes a high pressure which drives the heels into the floor.) The high pressures at the contact points cause the two surfaces to be cold-welded together which causes a further resistance to movement. Before one surface will slide over the other a force great enough to shear the cold welds must be applied. If the two surfaces are made from different metals particles of the softer metal will be transferred to the harder metal. Thus wear takes place on the surface of the softer metal.

Some examples of this are:

(1) A chip moving over the surface of a cutting tool will sometimes weld itself to the face of the tool. This condition is known as 'built-up edge'.

(2) When a shaft rotates in a bush, the bush becomes enlarged after a period and has to be renewed. Wear has taken place because of friction between the shaft and the bush.

The amount of frictional resistance. The amount of frictional resistance which occurs when we attempt to move one surface over another depends upon:

(1) The roughness of the two surfaces in contact. It is easier to move one polished steel block over another than it is to move a rough-cast block over another rough surface.

(2) The size of the force which keeps the two surfaces in contact. It is easier to move a light object over a rough

floor than it is to move a heavy object over the same floor.

Some of the advantages of friction. Friction is used to advantage in the workshop. Some ways are shown below:

1. Holding work in a vice (Fig. 2.58)

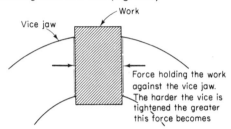

Fig. 2.58 The greater the force holding the work the greater the frictional resistance at the vice jaws. The larger the force, the more difficult it is for the work to slip from the vice.

2. Holding work in a lathe chuck (Fig. 2.59)

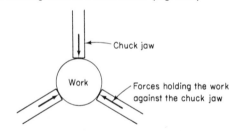

Fig. 2.59 The harder the chuck is tightened up the greater is the frictional resistance which holds the work in place against the forces which occur when cutting the workpiece.

3. Clamping work (Fig. 2.60)

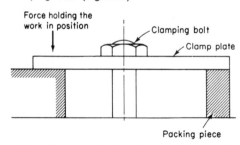

Fig. 2.60 The force holding the work in position is supplied by tightening up the clamping bolt (see Chapter 2.2 on Moments.) Friction between the work and the clamping plate holds the work in position against the cutting forces. The larger the force holding the clamp plate in position, the less chance there is that the work will slip.

4. Brakes (Fig. 2.61)

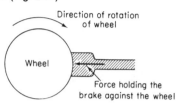

Fig. 2.61 Brakes are used to stop motion. The larger the force holding the brake against the wheel the greater is the frictional resistance. The more frictional resistance there is, the quicker the wheel stops. In order to increase the amount of friction the brake is fitted with material which has a high frictional resistance (e.g. Ferodo).

Disadvantages of friction. Sometimes friction causes serious disadvantages—some of which are given below:
1. Causing extra effort. Friction always tends to prevent movement of one part over another. Extra effort is therefore needed to cause movement. This is one reason why no mechanism or machine can be 100% efficient.
2. Wear. Friction causes wear between two moving surfaces. In some cases it can cause moving parts to seize up (see next section in this Chapter).
3. Heat generation. Friction often causes heat to be generated in moving parts. In many cases this heat has to be removed by using a cooling liquid.

The effect of speed. When two moving parts move slowly over each other not much heat is generated. However when the movement between two surfaces occurs at high speed a great deal of heat is generated (try moving your hand rapidly to and fro across a table). This heat is caused by the frictional resistance between the moving surfaces.

Heat causes the metals in contact to expand and it also increases the welding effect. Both of these contribute to increased wear. Sometimes seizure between the two surfaces in contact occurs. An examination of a seized shaft and bearing will show discoloration due to the heating effect. In addition the shaft will be roughened by particles of metal transferred from the bearing.

Machine slides usually operate at slow speeds. Hence only a small amount of wear takes place. Any wear that does occur results in the smoothing-off of the surfaces in contact.

Selection of materials to reduce friction. Generally speaking, surfaces which move over one another should be made from different materials. A steel shaft should rotate in a white metal or bronze bearing (a shaft should always be harder than the bearing). Bronze or brass is suitable for sliding on cast iron. However cast iron is frequently used in sliding contact with cast iron because of the particular properties of this metal.

Plastic materials such as nylon and P.T.F.E. are used for bearings because these materials give a low frictional resistance.

Lubrication

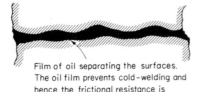

Film of oil separating the surfaces. The oil film prevents cold-welding and hence the frictional resistance is reduced. The resistance to motion is due only to the viscosity of the oil

Fig. 2.62   The effect of lubrication

The main purpose of lubrication is to reduce friction and wear (Fig. 2.62). The only resistance to motion is caused by the particles of oil sliding over one another. This resistance is called *viscosity* and it is a measure of the resistance of the oil to flow. Water has a low viscosity whilst treacle has high viscosity. Oils used as lubricants must have sufficient viscosity so that they are not squeezed out from between the two surfaces separated by the oil film.

There is another important property of lubricating oil —its oiliness. This is the ability of the oil to wet the metallic surfaces. Oils which contain molybdenum-disulphide are particularly good in this respect. Such oils reduce wear and they also prevent a breakdown of the oil film separating the metallic surfaces.

Lubrication of moving parts. Machine tools such as lathes, milling machines etc. can only behave satisfactorily if they are properly maintained and lubricated. Machine-tool lubrication involves slideway lubrication, spindle lubrication, head and gearbox lubrication and ball- and roller-bearing lubrication.

Slideway lubrication protects the slideways of the machine from wear. If they are properly lubricated with the correct kind of lubricant the movement of carriages, saddles, toolholders etc. will be precisely controlled. When unsuitable oil is used the movements of the carriage, saddle etc. will be jerky thus causing poor work to be produced.

Various methods are used to apply lubricant to the bearings of the workhead of a machine tool. Some are partly submerged in oil, some are lubricated by drop feed, some are periodically oiled by hand and some are lubricated by grease. When oil is used an oil of medium viscosity is adequate. For gearbox lubrication the oil must prevent wear occurring between the teeth of gears in contact. A medium oil is satisfactory for this purpose.

Ball and roller-bearings are used extensively in machine tools particularly for gearboxes, spindles and headstocks. The purpose of the lubricant is to reduce friction to a minimum and to protect the balls and rollers from corrosion and rust. Either grease or oil is used; oil is often preferable since it offers less resistance to the movement of the balls and rollers and it penetrates easily into small crevices and corners.

The manufacturers' recommendation should always be followed when using lubricating oils.

Exercise 2.5

1. What are the causes of friction?
2. Give two recognizable effects of friction.
3. What does the amount of frictional resistance depend upon?
4. By means of sketches show three ways in which friction is used to advantage in the workshop.
5. State three disadvantages of friction.
6. What effect does speed have when one surface is moving over another?
7. What is seizure? How does it occur?
8. State two different kinds of bearing combinations, naming suitable materials.
9. What is viscosity?
10. What is the purpose of lubrication?
11. What is oiliness and what effect does it have?
12. Which parts of machine tools are usually lubricated?

2.6 Work and power

Work. When an object moves because a force has been applied to it, the force has done work on the object. The amount of work done is found by using the formula:
work done =

force × distance moved in the direction of the force
That is, work done is the product of force and distance. The force is measured in newtons and the distance moved is measured in metres. Hence the unit of work done is newton metres (N m). However, in the SI units the joule (J) is used as the unit of work:

1 joule = 1 newton metre

Example. Find the amount of work done when a force of 50 newtons moves through a distance of 8 metres as shown in Fig. 2.63.

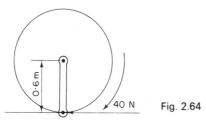

Fig. 2.63

Work done = force × distance moved
= 50 × 8 = 400 N m = 400 J

Example. Find the work done in lifting a mass of 100 kg through a height of 20 m.

To lift the object a force equal to the *weight* of the object must be applied. Hence,

Work done = weight of object × distance moved
= (100 × 9·81) × 20
= 981 × 20 = 19 620 N m
= 19 620 J or 19·62 kJ

When a handle of a machine or a shaft is rotated work is done which is again the product of force and distance.

Example. A force of 40 N is applied at the end of a handle 600 mm long. If the handle is given 7 complete revolutions, how much work is done?

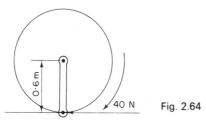

40 N   Fig. 2.64

When the handle moves through 1 revolution the force

has moved a distance equal to the circumference of a circle whose radius is 0·6 m.

∴ Distance moved by the force in 1 revolution
= 2 × π × 0·6 metres
Distance moved by the force in 7 revolutions
= 7 × 2 × π × 0·6 = 7 × 2 × $\frac{22}{7}$ × 0·6 = 26·4 m
Work done = force × distance moved
= 40 × 26·4 = 1056 J or 1·056 kJ

Power. Power is the rate of doing work:

$$power = \frac{work\ done}{time\ taken}$$

When the work done is measured in joules and the time taken is measured in seconds the power is measured in joules per second (J/s). However, in the SI units the watt (W) is used as the unit of power:

1 watt = 1 joule per second

Example. If 250 joules of work are used in 5 seconds, what is the power?

$$Power = \frac{work\ done}{time\ taken} = \frac{250}{5} = 50\ W$$

Example. A mass of 5000 kg is lifted through 8 m in 5 seconds. What power is needed?

Weight   = 5000 × 9·81 = 49 050 N
Work done = 49 050 × 8   = 392 400 J
Power   = $\frac{392\ 400}{5}$ = 78 480 W or 78·48 kW

A second way of calculating power is to use:
Power (in watts)

= Force (in newtons) × Speed (in metres per second)
This is a useful way of calculating the power used by a machine tool.

Example. The cutting force at the point of a lathe tool is 1200 N when the cutting speed is 18 metres per minute. Calculate the power used.

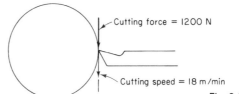

Fig. 2.65

Cutting speed = 18 m/min = $\frac{18}{60}$ = 0·3 m/s
Power = Force × Speed = 1200 × 0·3 = 360 W

Exercise 2.6

1. If a trolley is pulled with a force of 100 N for a distance of 8 m, how much work is done?
2. A force of 500 N moves an object a distance of 20 m in the direction of the force. How much work is done?
3. A mass of 200 kg is lifted through a height of 5 m. How much work is done?
4. A mass of 300 kg is lifted through 8 m. How much work is done?
5. A body having a weight of 350 N is lifted through 40 m. How much work is done?
6. A force of 80 N is applied to the end of a handle of a machine. The handle is 500 mm long and it makes 14 revolutions. How much work is done?
7. If 500 J of work are done in 10 seconds, what power is used?
8. An object is moved a distance of 20 m in 5 seconds. A force of 50 N is used. What power is used?
9. A body having a weight of 300 N is lifted through a height of 5 m in 2 seconds. How much power is needed?
10. A mass of 40 kg is lifted through a height of 10 m in 5 seconds. What power is used?
11. The cutting force at the point of a machine tool is 1500 N and the cutting speed is 18 m/min. What power is needed?
12. A brass bar is being machined at 90 m/min and the cutting force at the tool point is 600 N. What power is being used?

2.7   Simple chemistry

Oxidation. The atmosphere consists principally of nitrogen (78%) and oxygen (21%). Nitrogen is an inert gas which has little or no effect on most materials at ordinary temperatures. Oxygen, however, reacts with other elements to form oxides. A good example is the rusting of iron, the rust being iron oxide. Thus,

iron + oxygen → iron oxide

most metals rust in air due to the oxygen combining with the metal.

Combustion. We have seen that oxidation is the chemical combination of oxygen and an element or compound. When the element is iron, oxidation causes rust. When the compound is a fuel, the oxidation takes place very rapidly and large quantities of heat are freed. We call this form of oxidation 'combustion'.

Combustion is used as the source of heat in oxyacetylene welding. A mixture of oxygen and acetylene is used to provide the welding flame. A temperature of about 3200 °C may be attained in the hottest part of the flame.

Corrosion. The rusting of metals is more severe in damp atmospheres. For rusting or corrosion to take place both air and moisture are necessary. Air contains traces of water vapour and corrosive gases.

Severe rusting takes place on iron and steel but all metals rust to a certain extent. However, with metals like zinc, tin, chromium etc. the film of oxide which forms is very thin and dense. Once this thin film has formed no more rusting takes place.

Surface protection of iron and steel. Since corrosion destroys iron and steel, engineers must try to stop corrosion occurring. Some of the ways used are as follows:

1. *Cladding.* A thin sheet of corrosion-resisting metal is placed on the steel which is to be protected. Lead, zinc and copper are frequently used for this purpose.
2. *Galvanizing.* In this process the iron and steel is given a thin coating of zinc which protects it from corrosion. This zinc coating is frequently applied by immersing the article in a bath of molten zinc. This method is used for corrugated iron, window frames, buckets etc. An electrolytic method (see Chapter 2.10) is used when a thin coating of zinc is required. It is used when articles are to be later formed or soldered.
3. *Sherardizing.* When iron or steel is heated to a certain temperature in zinc dust, zinc is absorbed into the surface of the steel which then becomes rust proof. Nuts, bolts, screws, springs and chains are protected by this method.
4. *Metal spraying.* If metal powder is blown through a suitable flame the molten particles of metal will then stick to the steel to be protected. The coating produced by the spraying can be as thick as 0·25 mm if required

(much thicker than that produced by galvanizing or sherardizing). Aluminium and zinc are used in this way to protect iron and steel.

5. *Painting.* Protection against corrosion may be obtained by painting the surface to exclude air and moisture. Before painting, all traces of rust must be removed otherwise the rusting will continue under the paint. The paint may be applied by brushing, spraying or dipping the article.

6. *Anodizing.* Anodizing is used to provide a decorative and corrosion-resistant coating to aluminium alloys. In anodizing, the aim is to thicken the oxide film which forms naturally on the surface of the aluminium alloy. Hard anodizing is used to give wear resistance.

7. *Electroplating.* In electroplating, a thin layer of a rarer metal (e.g. nickel, chromium, tin, etc.) is deposited on a baser metal (e.g. iron and steel) to provide a coating which prevents corrosion of the baser metal. The method used to do this is called electrolysis (see Chapter 2.10). The table opposite shows the more common forms of electroplating.

Fig. 2.66 shows the main methods used in the avoidance of corrosion.

Reduction. It is possible to remove oxygen from a compound. The chemical reaction is then the opposite to oxidation and we call it *reduction.* During certain heating processes the surface of steel is often discoloured owing to oxidation. For some purposes this is undesirable and we then heat the steel in a reducing atmosphere, because this reducing atmosphere prevents oxidation. Some oxygen is also removed from the surface of the steel. Thus, the steel heated in a reducing atmosphere remains bright and clean.

Fluxes, such as killed spirits (see Chapter 4.11) remove oxygen from the surface of the work and hence makes soldering easy.

Exercise 2.7

1. What is oxidation and what is its effect on metals?
2. What is combustion? Give two workshop applications.
3. What is corrosion?
4. Why do metals like zinc and tin not corrode?
5. Name three metals which are used for cladding.
6. What is galvanizing?
7. What is anodizing?
8. Name two metals used when electroplating.
9. How is electroplating done?
10. What is reduction? Give three examples where reduction is of value in engineering.

ELECTROPLATING

Metal	Uses	Remarks
Cadmium	Corrosion protection of iron and steel.	Use mainly where thin coatings are needed.
Nickel	Sometimes used as a decorative coating but is used more frequently as an undercoating for chromium.	
Chromium	Used for decorative coatings. Hard chromium is used where wear resistance is required.	Has a high resistance to wear and tarnishing. It is usually plated over nickel which protects against corrosion.
Tin	Corrosive-resistant coating particularly in the food industry.	
Zinc	Corrosion protection of iron and steel.	

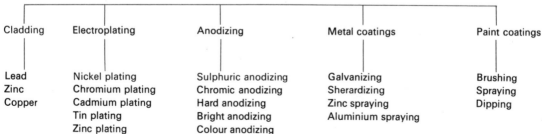

METHODS USED IN AVOIDANCE OF CORROSION

Cladding	Electroplating	Anodizing	Metal coatings	Paint coatings
Lead Zinc Copper	Nickel plating Chromium plating Cadmium plating Tin plating Zinc plating	Sulphuric anodizing Chromic anodizing Hard anodizing Bright anodizing Colour anodizing	Galvanizing Sherardizing Zinc spraying Aluminium spraying	Brushing Spraying Dipping

Fig. 2.66

2.8   Materials used in engineering

Properties of materials. All engineering materials possess certain properties. These properties are given special names as follows:

Ductility. Metals which can be drawn out into a fine wire are said to be ductile. Wrought iron and mild steel are examples.

Malleability. This is a property very similar to ductility. A malleable material can be beaten or rolled into plates without cracking. Lead is a typical example and it can be beaten out when cold. Other metals, when they are hot, can be greatly distorted by hammer blows without cracking. Mild steel and wrough iron are two such metals.

Strength. A strong metal can withstand larger forces than weak metals. The ultimate tensile stress (see Chapter 2.1) is a measure of the strength of a material. Steel is a strong metal but aluminium is comparatively weak.

Elasticity. This is the ability of a material to return to its original dimensions after suffering deformation resulting from an applied force. Some materials, rubber is an example, are very elastic but others, such as cast iron, possess very little elasticity.

Toughness. Some materials can stand repeated blows with a hammer without breaking. These materials are said to be tough. Other materials break easily when struck with a hammer. These metals are said to be brittle. Cast iron is a brittle material but mild steel is tough.

Hardness. Hard materials are difficult to file. They are also difficult to mark with a centre punch. Soft materials, on the other hand, are easy to mark with a centre punch.

Electrical conductivity. Some materials easily allow the passage of an electric current and their electrical conductivity is said to be high. Copper and aluminium possess high electrical conductivity. On the other hand, some materials such as glass, ebonite and porcelain are poor conductors of electricity. Such materials are called insulators (see Chapter 2.10).

Heat conductivity. Some materials conduct heat very well. If one end of a copper bar is held in a fire the other end rapidly becomes warm, because the copper has conducted the heat from the hot end to the cold end. Most metals are good conductors of heat, although some are better than others. Materials such as wood, paper and glass are poor conductors of heat and such materials are often used as heat insulators.

Ferrous metals. The metals used in industry fall into two groups—those that contain iron and those that do not. Metals which contain iron are known as *ferrous metals*. Those that do not are called *non-ferrous metals*. The chief ferrous metals are cast iron and steel whilst lead, copper, aluminium etc. are non-ferrous metals.

Steel. Steel is an alloy of carbon and iron although other elements may be present. There are two main kinds of steel—the plain carbon steels and alloy steels. The plain carbon steels contain only iron and carbon but the alloy steels contain other elements as well. At this stage, however, only the plain carbon steels will be discussed.

The effect of carbon on the properties of steel. All metals possess certain properties. They may be hard or soft, strong or weak, tough or brittle. With the plain carbon steels it is the amount of carbon that determines which of these properties the steel possesses. The amount of carbon used is very small and varies from 0·1% to 1·4%. Wrought iron contains no carbon and the metal is soft, ductile and fairly weak. As carbon is added the steel becomes progressively harder, stronger and more brittle. A steel containing 0·25% carbon is comparatively soft, weak and tough but a steel containing 0·8% carbon is hard, strong and fairly brittle.

Uses and properties of plain carbon steels. The table below gives an indication of the carbon content and uses of plain carbon steels.

PLAIN CARBON STEELS

Name of steel	Carbon %	Uses
Dead mild	0·1 to 0·13	Thin sheets.
Mild	0·15 to 0·3	Forgings, general workshop purposes, girders.
Medium carbon	0·3 to 0·5	Forgings.
	0·5 to 0·7	Hammers and snaps for riveting.
High carbon	0·7 to 0·9	Springs, shear blades.
	0·9 to 1·1	Cold chisels, punches screwing dies.
	1·1 to 1·4	Hand files, gauges and metal cutting tools.

Steel may be obtained in a variety of forms such as plates, thin sheet, billets for making forgings, bars, strip and wire.

Cast iron. *Grey cast iron* contains about 3·5% carbon. It is moderately hard and very brittle. It can easily be broken by a hammer blow and thin castings may break if they are dropped. Although grey cast iron is brittle it is very rigid and strong in compression. It is used for the beds of machines, surface plates and vice bodies.

The brittleness of grey cast iron stops this material being used where shock loads are likely to be encountered. *Malleable cast iron* partly overcomes this disadvantage and such cast irons are used for making thin castings particularly where shock loads are likely.

Other types of cast iron such as *spheroidal graphite* are used where high strength is needed.

Non-ferrous metals. The properties and uses of the more common non-ferrous metals are shown in the following table:

NON-FERROUS METALS

Metal	Properties	Uses
Copper	Malleable, ductile, good conductor of heat and electricity, resists corrosion.	Castings, forgings, wire, sheet. Used for water pipes, electric cables, soldering irons, etc.
Tin	Malleable, ductile, fairly weak.	Used for coating thin steel sheets to make tinplate.
Zinc	Resists corrosion.	Used for making galvanized sheet.
Aluminium	Light, weak, soft, ductile and malleable. Resists corrosion.	Generally only used in aluminium alloys, and for electric cables.

Non-ferrous alloys

Brass is made by alloying copper and zinc. Some typical brasses are given in the table in the next column:

The addition of lead to brass improves its machining properties. A typical leaded brass is 57% copper, 41% zinc and 2% lead.

Composition		Properties and uses
Copper	Zinc	
70%	30%	Very strong and ductile. Used for tubes, wire, pressings and cartridge cases.
63%	37%	Ductile and strong. Used for cold pressings.
60%	40%	Muntz metal. Used for castings, water fittings, etc.
50%	50%	Used for brazing as it has a low melting point.

Light alloys. Pure aluminium is weak and soft but when it is alloyed with small amounts of other metals its properties are greatly improved. The resulting alloys are called light alloys and these are used extensively where least weight is important (e.g. in the aircraft industry).

Duralumin is a light alloy of aluminium with about 4% copper and small amounts of manganese and magnesium. It can be obtained in sheets, bars and tubes and it can be forged. It is strong, its strength being nearly equal to that of mild steel. It is used extensively in the aircraft industry. it is possible to obtain a light alloy for almost all purposes and one suitable for casting has about 12% silicon and 88% aluminium.

White bearing metals are composed of tin, antimony and copper or tin, antimony and lead in various amounts according to the work the bearing has to do. White-metal bearings are used in motor and aero engines and in electrical machines.

Bronze. When alloyed with tin, copper forms an alloy called bronze. The best known is *gunmetal* which contains 88% copper, 10% tin and 2% zinc. Bronze is used mainly for castings which must be corrosion resistant.

Cutting tool materials. When a tool cuts, the chip passes over the face of the tool. This action causes heat to be generated and the tool becomes hot. The amount of heat generated increases with the cutting speed. The rapid removal of metal demands high cutting speeds and so the material of the cutting tool must keep its hardness at high temperatures. Plain carbon steel tools become too soft to cut at about 250 °C and because of this other materials are used when high cutting speeds are needed. The table overleaf gives a list of cutting tool materials and their uses, etc.

CUTTING TOOL MATERIALS

Material	Uses	Remarks
Carbon tool steel	Large taps, reamers and dies.	Unsuitable for high cutting speeds as it softens at temperatures above 250 °C.
High-speed steel	Drills, turning and shaping tools, milling cutters, etc.	Retains hardness at temperatures up to 650 °C. High cutting speeds can be used. These steels contain 14–22% tungsten, and can be used for complete cutters or as tool bits welded to carbon-steel shanks.
Stellite	Bits for turning tools and milling cutters.	This material is extremely brittle and can be used only for tool bits. Keeps its hardness at high temperatures.
Tungsten carbide	Bits for turning tools and milling cutters.	Very brittle and used only for tool bits brazed to carbon-steel shanks.
Ceramics	Bits for turning tools and milling cutters.	These consist principally of aluminium oxide and they have a high hardness at elevated temperatures. Very brittle.

Plastic materials. These can be broadly divided into two groups:

(1) Thermo-softening or thermoplastic materials which can be softened and resoftened repeatedly by heating them to a required temperature. Thus an object made from a thermoplastic material can be reshaped if desired by applying heat.

(2) Thermo-setting materials which undergo a chemical change during the initial process of being shaped. These materials cannot be resoftened and hence their shape is permanent.

The tables following give the properties and uses of the most common plastic materials.

THERMO-SETTING PLASTICS

Name	Properties	Uses
Bakelite	Good insulator. Strong in compression. Some types are brittle.	Containers of various kinds.
Silicones	Water-repellent and can withstand high temperatures.	High temperature electrical insulation work. Additives to oils, waxes and rubbers.
Epoxy resins (Araldite)	Resistant to heat and moisture. Good electrical properties.	Sealing of electrical components against moisture etc. Used as a basis of adhesives.

THERMOPLASTICS

Name	Properties	Uses
Cellulose acetate	Tough, reasonably inflammable, good electrical insulator. Has high water absorption. Clear.	Welding goggles, machine guards and welding guards.
Polyvinyl chloride (PVC)	Good thermal insulator. Softens at about 70 °C. Tough and rubber-like and non-inflammable.	Insulation of electric cables, plating baths, factory ducting and chutes.
Perspex	Clear and lighter than glass. Rather brittle, good insulator and unaffected by dilute acids and concentrated alkalis.	Windscreens, roof lights, machine guards, jigs and models.
PTFE	Good insulator and very low coefficient of friction. Weak. May be used at any temperature between −250 °C and 250 °C.	Gaskets, packings, filters, seals, insulating tapes, electrical insulators.
Nylon	High strength, high melting point, high resistance to wear and low coefficient of friction. May be used at temperatures between 50 °C and 150 °C.	Gears, bushes and bearings for light engineering jobs. Washers, wheels and rollers and electrical insulators.
Polythene	Chemically inert to most liquids. Tough and slightly elastic. Good insulator.	Moulded containers and electrical insulation.

Exercise 2.8

1. Which of the following metals are ductile:
   cast iron, mild steel, copper, zinc, lead?
2. Which of the following metals are malleable:
   lead, copper, cast iron, mild steel?
3. Place the following metals in their order of strength:
   aluminium, mild steel, lead, wrought iron.
4. What is elasticity? Give two examples of metals which possess this property.
5. What is the difference between toughness and brittleness?
6. Place the following metals in order of hardness:
   cast iron, high carbon steel, mild steel and wrought iron.
7. Name two metals which are good heat conductors.
8. Name four materials which can be used as insulators against heat.
9. Name four non-ferrous metals.
10. What is steel?
11. Arrange the following metals in their order of brittleness. Place the most brittle first: mild steel, wrought iron, medium carbon steel, cast iron.
12. What effect does carbon have on the properties of steel?
13. Grey cast iron is a brittle metal. What other types of cast iron are available which are less brittle?
14. Why is aluminium seldom used in engineering unless it is alloyed with another metal? What metals are used?
15. What effect does lead have on brass?
16. State three physical properties of aluminium that makes its alloys valuable engineering metals.
17. Why has plain carbon steel been superseded by other cutting materials?
18. Name four cutting materials used in the machine shop and give two uses for each material.
19. Give two examples of cutting tool materials which are used only as tool bits. Why are they used in this way?
20. Plastic materials can be divided into two groups. Name these groups and state the differences between them.
21. Name plastics that would be suitable for the following purposes:
    (a) welding goggles   (b) aircraft windscreens
    (c) electrical insulation   (d) acid containers
    (e) gears   (f) bearings
    (g) seals   (h) bushes
22. State the most important properties of:
    (a) Perspex   (b) Cellulose acetate
    (c) PVC   (d) Nylon
    (e) Polythene   (f) Bakelite
23. Where has polythene replaced rubber?
24. What are epoxy resins used for?

2.9 Metallurgy

Crystals in metals. Many chemical substances are naturally crystalline. For instance cane sugar and common salt are crystalline and the grains of each substance have the same shape. In the case of salt this shape is a perfect cube.

Crystals may be formed in one of two ways:
(1) By cooling a substance, which has been molten, until it solidifies.
(2) By allowing a substance, which has been dissolved in some other substance, to come out of solution (e.g. evaporating salt solution).

All metals are made up of a mass of crystals. When molten steel (or any other metal) cools down it starts to crystallize out. Naturally, the parts of the metal to solidify first will be those that are the coolest. Fig. 2.67 represents

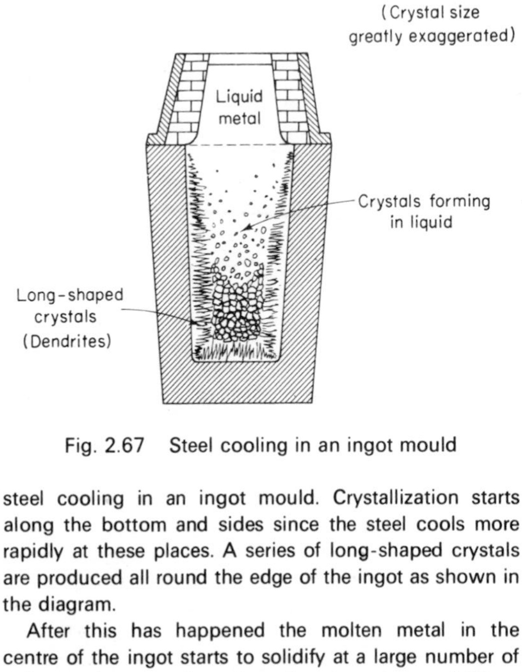

(Crystal size greatly exaggerated)

Fig. 2.67   Steel cooling in an ingot mould

steel cooling in an ingot mould. Crystallization starts along the bottom and sides since the steel cools more rapidly at these places. A series of long-shaped crystals are produced all round the edge of the ingot as shown in the diagram.

After this has happened the molten metal in the centre of the ingot starts to solidify at a large number of points at the same time. At each point, where solidification takes place, a crystal starts to grow outwards in all directions. The crystal continues to grow as long as there is molten metal left to feed the crystal and until it meets the surfaces of neighbouring crystals growing in the same way. Crystal growth then stops.

As shown in Fig. 2.68 the steel, when it has solidified completely is made up of a mass of crystals. Every crystal, except those on the outside is completely surrounded by neighbouring crystals and separated from them by crystal boundaries or grain boundaries.

Fig. 2.68   The crystals in a steel ingot

Every crystal is weaker in one particular direction than it is in other directions. If a piece of steel was made of one crystal only, this would be a very serious source of weakness. At a comparatively low stress in the direction of the weakness the metal would distort and finally break. But because the steel is normally made up of a large number of crystals arranged in a 'higgledy-piggledy' manner the direction of the weakness in one crystal is different from that of its immediate neighbours. Thus when a force is applied there are only a few crystals which are ready to start giving way at the same time and these crystals will be widely scattered throughout the metal. To a large extent they will be prevented from breaking by the crystals surrounding them.

We may now appreciate the importance of crystal size in determining the strength and toughness of a metal. When the individual crystals are large, more damage is done to the metal when that crystal breaks than would happen if the crystals were small. Thus the crystal size in the metal should be as small as possible. Toughness in a metal can be obtained by treating it in such a way that the size of the crystals is reduced.

Recrystallization. When a piece of lead is squeezed between a pair of rolls (Fig. 2.69 (a)) the act of squeezing it causes the original crystal structure to break down and be replaced by a fresh one. The lead has recrystallized itself. For this reason lead remains soft after squeezing, rolling or bending it. There is no need to heat the lead—it will recrystallize when worked in the cold state.

Steel and other metals, however, will not recrystallize in the cold state (Fig. 2.69(b)). The action of rolling them causes the crystals to elongate as shown in the

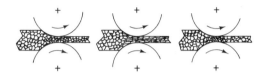

(a) Cold–rolling lead
(recrystallization)

(b) Cold–rolling steel
(crystal deformation)

(c) Hot–rolling steel
(recrystallization)

Fig. 2.69   Effects of hot and cold rolling

diagram. An elongated and distorted crystal structure of this kind makes the metal harder and stronger so that it is more resistant to further cold work. We say that the metal has been 'work hardened'.

After cold work, however, the metal can be made to recrystallize under the action of heat alone. When the crystal structure of the metal has been distorted, heating it puts the metal back into its original state of softness and ductility by causing a fresh crystal structure to be formed (Fig. 2.69 (c)). When drawing wire by pulling it through a succession of smaller and smaller holes to reduce its diameter, the wire finally becomes so hard and brittle that further drawing is impossible. The wire must be heated to cause it to recrystallize and hence soften. Sometimes we can combine these two operations and bring about recrystallization by hot working. Hot rolling of steel has the same effect on the crystal structure as the cold rolling of lead.

Heating and cooling steel. We have seen that when steel is heated it recrystallizes. If the steel is heated to too high a temperature the crystal size will increase and go on increasing (until it becomes too large) as the steel cools down. At too low a temperature the steel will not recrystallize properly. Therefore, when heating the steel we must pay particular attention to the temperature used.

For reasons already discussed a large grain (or crystal) size gives a soft, weak steel whilst a small (or fine) grain size gives a harder, stronger metal. Thus the correct type of crystalline structure plays a very important part in determining the strength and toughness of the steel.

It is the rate at which the steel is cooled which determines the final crystalline structure of the steel. A very slow cooling rate produces large crystals and hence a soft, weak metal. A fast cooling rate produces small crystals and hence a hard, strong metal.

Heat treatment of steel. Sometimes we need to soften steel so that it becomes easier to work. At other times we may need to harden steel so that it resists wear

or so that it can cut other metals. These properties can be given to a steel by suitable heat treatment.

For the present, we will consider the heat treatments of annealing, normalizing, hardening, and tempering.

Annealing. This is done to soften the steel so that it becomes easier to work. The steel is heated to the required temperature and it is held at this temperature for a short time. It is then cooled very slowly. If the heat treatment takes place in a furnace (as is usually the case) the cooling may be done by leaving the work in the furnace and cooling both furnace and work together. If this is not possible, the work may be cooled in sand or ashes, as these materials prevent the heat from escaping too quickly. The temperature to which the work is heated varies with the carbon content of the steel. The table below gives suitable temperatures for various steels.

ANNEALING TEMPERATURES

Type of steel	Carbon content	Annealing temp. °C
Dead mild	less than 0·12%	875–925
Mild	0·12–0·25%	840–870
Medium carbon	0·25–0·5%	815–840
	0·5–0·9%	780–810
High carbon	0·9–1·3%	760–780

Normalizing. After steel has been worked it may not be as strong and as hard as it should be and it may also be strained (i.e. have distorted crystals). To bring the metal to its best condition, for use in service, it is normalized. The steel is heated to the annealing temperature and held at this temperature for a short time. It is then cooled in still air. Thus the cooling rate is much quicker for normalizing than it is when annealing. The effect is to make the crystal size smaller, thus increasing the hardness and toughness of the steel.

Hardening. To harden steel we heat it to the annealing temperature and quench it in water or oil. Thus the cooling rate is very quick and a very fine grain structure is obtained. Water causes the steel to be harder than does oil, but with some jobs, quenching in water may cause the work to distort and crack. Steels containing less than 0·3% carbon are seldom hardened because the amount of hardening obtained is too small to be of any account.

Tempering. A steel which has been hardened is very brittle and it will crack and chip when used. This is particularly true of cutting tools. Tempering, i.e. reheating the steel and then quenching it, takes away some of this brittleness and makes the tool tougher. The temperature to which the steel is heated depends upon the use to which the article is to be put. This temperature is always lower than the annealing temperature. It should be noted that tempering causes the hardened steel to become softer—the higher the tempering temperature the softer the steel will be. The tempering temperature is sometimes judged by the temper colour (see table below) which appears on the freshly polished surface of steel when heated. This temper colour appears because the thin oxide film on the surface of the steel changes colour. It is not a heat colour in the true sense.

TEMPERING TEMPERATURES

Article	Temperature °C	Temper colour
Turning and shaping tools	230	Pale straw
Milling cutters and drills	240	Dark straw
Taps	250	Brown
Twist drills and reamers	260	Brownish-purple
Cold chisels	280	Purple
Springs	300	Blue
Toughening for constructional steels	450–650	—

Heat treatment of aluminium and its alloys. When aluminium is cold-worked by rolling, drawing and pressing, the hardness and strength of the metal is increased but the ductility is decreased. In order to continue working, softening (by annealing) is sometimes necessary. The metal is heated to about 350 °C so as to cause recrystallization.

To obtain the highest strength the metal is heated to about 500 °C and quenched in water. The metal then becomes quite soft and ductile but only for a relatively short time. If the alloy is left at room temperature it starts to become harder and stronger until after about five days the metal reaches maximum strength and hardness. This process is called *age-hardening*. Age-hardening can be delayed if necessary by keeping the components cold in a refrigerator, a technique often applied to rivets for the aircraft industry.

Heat treatment of copper. Copper may be annealed (softened) by heating it to 320 °C. After heating the copper may be quenched in water or it may be allowed to cool naturally.

Copper cannot be hardened by heat treatment, but the hardness desired can be obtained by cold working. The hardness of copper is often quoted in tempers. Soft temper refers to annealed copper, half-hard temper and hard temper refer to partially and fully work-hardened metal respectively.

Exercise 2.9

1. Describe two ways by which crystals can be formed.
2. Why is crystal size important in determining the strength and toughness of a metal?
3. What happens to lead when it is squeezed between a set of rolls?
4. What is recrystallization? How may steel be recrystallized?
5. What determines the final crystal size in steel after it has been heated?
6. Describe the heat treatment necessary to make steel soft enough for cold working.
7. Why is normalizing necessary?
8. In making a bracket from mild-steel thin sheet it is found that it is difficult to bend the flanges. What heat treatment would you recommend? How would you do this heat treatment?
9. What is the purpose of tempering?
10. Describe the heat treatment of a cold chisel made from 0·8% carbon steel.
11. What is age-hardening of aluminium and its alloys?
12. How is copper annealed?

2.10   Basic electricity

The structure of the atom. If any material is split up into small pieces and each piece is split again and again we shall eventually obtain the smallest possible piece of the material. These extremely small pieces are called *atoms* and all materials consist of millions upon millions of them.

An atom itself may be split up into several even smaller particles. Fig. 2.70 shows the structure of the hydrogen atom. We see that it is like a sun with one planet spinning around it. The planet is called an *electron* and the sun is called the *nucleus*.

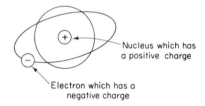

Fig. 2.70   The structure of the hydrogen atom

All atoms consist of electrons spinning around a nucleus. There are different types of atoms and different types of nucleii but all electrons are exactly the same. Fig. 2.71 shows an aluminium atom which is entirely different from the hydrogen atom.

The atoms of various materials are different because they contain different numbers of electrons.

A hydrogen atom has 1 electron.
A carbon atom has 6 electrons.
An aluminium atom has 13 electrons.
A copper atom has 29 electrons.

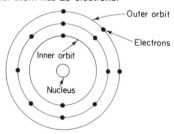

Fig. 2.71   The structure of the aluminium atom

Electrons circle the nucleus in definite orbits but some orbits are further from the nucleus than others as shown in Fig. 2.71. Electrons have a negative charge of electricity whilst the nucleus has a positive charge. In any atom the number of negatively-charged electrons is exactly equalled by the number of positive charges in the nucleus. Now unlike charges attract each other whilst like charges repel each other (Fig. 2.72). The nearer the charges are to each other the greater the degree of attraction or repulsion. Hence the electrons in the outer orbits are less strongly attracted to the nucleus than those in the inner orbits and so are more easily moved from their orbital positions.

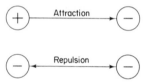

Fig. 2.72   Like charges repel whilst unlike charges attract

Those electrons which are able to free themselves are known as loosely-bound or free electrons. Loosely-bound electrons are able to leave one atom and join another without difficulty. Normally the loosely-bound electrons move in all directions in a haphazard fashion and the overall effect is that these movements cancel one another out (Fig. 2.73).

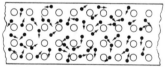

Loosely bound electrons move in a haphazard fashion. There is no flow of electricity

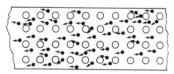

When the movement of the free electrons takes place in the same direction a flow of electrons (i.e. a flow of electricity) occurs. The current is measured by counting the number of electrons that flow through the material in each second

Fig. 2.73   How an electric current occurs

However if the movement of the free electrons takes place in the *same direction* we obtain a flow of electricity. To get the free electrons to move in one direction we need some form of energy. In practice chemical or magnetic energy is used.

Quantity of electricity. The smallest known charge of electricity is the negative charge on the electron. These charges are exceedingly small and the practical unit of quantity of electricity is the *coulomb* which is equal to 6 289 000 000 000 000 000 electrons.

How chemical action produces a flow of electricity. The primary cell is used to produce electricity. This consists of two plates, one copper and the other zinc immersed in acid such as dilute sulphuric (Fig. 2.74). The liquid in which the plates are immersed is called an electrolyte.

If we could see the cell working we would see that the electrolyte removes electrons from the zinc plate and pushes them onto the copper plate. Thus the zinc plate loses electrons and becomes positively charged, whilst the copper plate gains electrons and becomes negatively charged. The copper plate is called the negative terminal, whilst the zinc plate is called the positive terminal.

Now let us connect a wire between the two terminals. The positive terminal has a shortage of electrons and there is thus a strong attraction between the terminal and electrons from nearby atoms. These nearby atoms then lose electrons and attract electrons from neighbouring atoms and so a flow of electricity is created.

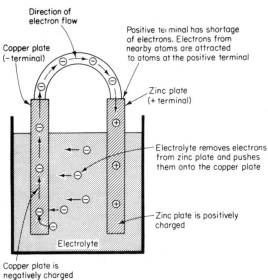

Fig. 2.74   How chemical action causes a flow of electricity

How magnetism produces electricity. Nearly all the electricity used is produced by generators in power stations. The generator may be driven by steam or water power but no matter how it is driven the electricity is produced as a result of magnetism.

When a wire is moved past a bar magnet an electric current is produced. We can see that this is so by connecting the wire across a sensitive meter (Fig. 2.75). When the wire is moved past the magnet the needle of the meter deflects, showing that an electric current has been produced. When the wire and magnet are stationary no current is produced.

In order to produce a continuous movement the wire is moved with a circular motion. This method of producing electricity by means of a wire travelling in a circle past a stationary magnet is the principle used in the electric generator.

It has been found that the stronger the magnet and the greater the length of wire passing across the magnet the greater will be the flow of electricity. The wire is therefore made into a coil and the coil is moved past the magnet very quickly. A diagram of a simple generator is shown in Fig. 2.76. The output of a generator is usually controlled by altering either the strength of the magnet or the speed of the coil.

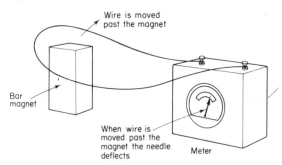

Fig. 2.75   Electricity produced by magnetism

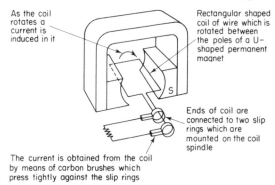

Fig. 2.76   Simple generator

Conductors and insulators. Materials with loosely-bound electrons are good conductors of electricity. In most metals the loss of electrons from atoms readily occurs and hence an electric current may be passed through these materials The best conductors are copper, silver and aluminium.

Some materials such as glass, porcelain, plastics, wood and paper have a very strong bond between their nucleus and the outer electrons. It is very difficult to get the electrons to leave the atoms and a very strong electrical field is needed before they will. Such materials are insulating materials and it is very difficult to pass an electric current through them (see Fig. 2.77).

The measurement of current. An electric current is caused by a flow of electrons through the material. It is measured by finding the number of electrons that flow through the material per second (Fig. 2.73). We have seen that the practical measure of quantity of electricity is the coulomb. If 1 coulomb of electricity flows across a given section of the material in 1 second, the current is 1 *ampere* (A). If 8 coulombs flow in 2 seconds the current is 4 coulombs per second or 4 amperes. Hence,

quantity of electricity (in coulombs)

= current flow (in amperes) × time (in seconds)

Example. A current of 10 A flows for 20 seconds. Find the quantity of electricity used.

Quantity = 10×20 = 200 coulombs

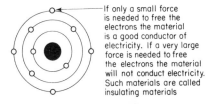
If only a small force is needed to free the electrons the material is a good conductor of electricity. If a very large force is needed to free the electrons the material will not conduct electricity. Such materials are called insulating materials

Fig. 2.77   The reason why some materials conduct electricity and others do not

Electromotive force (e.m.f.). The free electrons in a conductor have to be forced through the stationary atoms before an electric current can occur. Just as hydraulic pressure forces water through a pipe so an electrical pressure is needed to force the electrons through a wire. A water pump provides the necessary hydraulic pressure, whilst a battery or a dynamo provides the electrical pressure. The force which the battery or

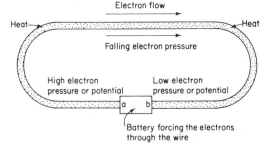

Electron flow

Heat→   ←Heat

Falling electron pressure

High electron   Low electron
pressure or potential   pressure or potential

a b

Battery forcing the electrons through the wire

Fig. 2.78   The idea of e.m.f.

dynamo applies to the electrons is called the electromotive force (e.m.f.). The way in which this is done is shown in Fig. 2.78.

The battery (or dynamo) takes in electrons at low potential (or pressure) and delivers them at a higher potential. This increase in potential is the e.m.f. of the battery. The work which is done by the potential difference in causing a steady flow of electrons to flow through the wire is converted into heat or work. Potential difference is measured in *volts* (V).

The potential difference between a and b (Fig. 2.78) is defined by:

$$E \text{ (volts)} = \frac{\text{heat generated in joules per second}}{\text{current in amperes}}$$

Electrical Power. Electrical power is measured in watts (W).

1 watt   = 1 joule per second

$$E \text{ (volts)} = \frac{P \text{ (power in watts)}}{I \text{ (current in amperes)}}$$

$$P \text{ (watts)} = E \text{ (volts)} \times I \text{ (amperes)}$$

Example. An electric fire is designed to run at 240 V and and 15 A. How much power is used?

Power = 240×15 = <u>3600 W</u> or <u>3·6 kW</u>

Example. A 2-kW electric fire is designed to run at 240 V. What current supply is needed?

$$I \text{ (amperes)} = \frac{P \text{ (watts)}}{E \text{ (volts)}} = \frac{2000}{240} = \underline{8·33 \text{ A}}$$

Electrical energy. It was shown in Chapter 2.6 that the unit of work is the joule. The same unit is used for electrical energy because work must be done in order to cause a flow of electricity. However, the joule is a very small amount of electrical energy and in practice the *kilowatt hour* (kW h) is used. When 1 kW is used for 1 hour the amount of electrical energy used is 1 kilowatt hour.

Example. An electric fire is rated at 2 kW. If it is used for 3 hours how much electrical energy in kilowatt-hours is consumed?

Electrical energy consumed

= number of kilowatts × time

= 2×3 = <u>6 kW h</u>

Example. A heating element takes a current of 4 A from a 250-V supply. If the element is used for 3 hours how much electrical energy (in kWh) is consumed?

P (watts)   = E (volts) ×I (amperes)

                 = 250×4 = 1000 watts = 1 kW

Energy consumed = kilowatts × time = 1 ×3 = <u>3 kW h</u>

The kilowatt hour is sometimes referred to as the 'Board of Trade Unit' but more often it is simply called the 'unit' of energy. Thus, when we say that 500 units of electricity have been used we mean that 500 kW h have been consumed. Electricity bills are calculated on the number of kilowatt hours used. Thus, if the price per unit is 2p then, if 500 units are used the cost will be 2×500 = 1000p = £10·00.

Electrical resistance. Some materials give up their loosely-bound electrons very easily and these materials offer very little resistance to current flow. Other materials, however, hold on to their outer electrons. Such materials offer considerable opposition or resistance to current flow. Every material offers some resistance to current flow whether large or small.

The factors which control resistance are:
1. *Material.* Some materials offer more resistance to current flow than others.
2. *Length.* The longer the length of a conductor the greater the resistance.
3. *Cross-sectional area.* The greater the cross-sectional area the smaller the resistance.
4. *Temperature.* For most materials the resistance increases as the temperature increases.

The unit of resistance is the ohm (Ω).

Ohm's law. The relationship between volts, amperes and ohms is given by the following formula:

$$E \text{ (volts)} = I \text{ (amperes)} \times R \text{ (ohms)}$$

This equation is known as *Ohm's Law.* The relationship may be stated in two other ways:

$$I \text{ (amperes)} = \frac{E \text{ (volts)}}{R \text{ (ohms)}}$$

$$R \text{ (ohms)} = \frac{E \text{ (volts)}}{I \text{ (amperes)}}$$

Examples
1. A 240-V supply is wired to a 60-ohm resistor. What current is flowing?

$$I \text{ (amperes)} = \frac{E \text{ (volts)}}{R \text{ (ohms)}} = \frac{240}{60} = 4$$

The current flowing is <u>4 A</u>
2. A current of 4 amperes is flowing through a resistance of 50 ohms. What voltage is needed?

$$E \text{ (volts)} = I \text{ (amperes)} \times R \text{ (ohms)} = 4 \times 50 = 200$$

The voltage required is <u>200 V</u>
3. A 240-V supply produces a current of 6 A. What is the resistance of the conductor?

$$R \text{ (ohms)} = \frac{E \text{ (volts)}}{I \text{ (amperes)}} = \frac{240}{6} = 40$$

The resistance of the conductor is <u>40 Ω</u>

The electric circuit. An electric circuit is a complete electrical pathway in which the current can flow from the negative terminal to the positive terminal. A lamp connected across a battery forms a simple electric circuit (Fig. 2.79 (a)). So long as the pathway is unbroken, current flows and we have a closed circuit. If the pathway is broken at any point an open circuit results and the current ceases to flow.

Thus, current flows until the circuit is broken. We may stop the current flow in the lamp circuit by removing the lead from the battery or by using a switch (Fig. 2.79 (b)). The electrical circuit symbols are shown in Chapter 3.8 of this book.

(a) A closed circuit consisting of a filament lamp connected across the terminals of a cell

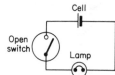

(b) An open circuit. The same circuit as above but with a switch added. The switch is open so the circuit is open

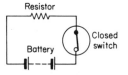

(c) A closed circuit consisting of a battery, switch and resistor. The switch is closed so the circuit is closed

Fig. 2.79   Simple electrical circuits

Resistances in series. Fig. 2.80 shows three resistances joined together, end to end, in one long line. When resistors are connected in this way they are said to be connected in series.

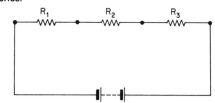

Fig. 2.80   Resistances in series

In a series circuit the current must flow through each resistor in turn. This means that everywhere in the circuit the current must be the same. If one switch is placed in the circuit (Fig. 2.81) then this switch controls the whole circuit and it does not matter where it is placed.

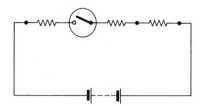

Fig. 2.81   One switch in a series circuit controls the entire circuit

Resistances joined in series have a total resistance equal to the total sum of the resistances.

Total resistance = resistance of resistor 1
+ resistance of resistor 2
+ resistance of resistor 3+ . . .

$$R_T = R_1 + R_2 + R_3 + \ldots$$

Examples

1. In Fig. 2.81 the resistances are 30 Ω, 20 Ω and 50 Ω respectively. What is the total resistance?

$$R_T = R_1 + R_2 + R_3 = 30 + 20 + 50 = \underline{100\ \Omega}$$

2. Find the current flowing in the circuit which is shown in Fig. 2.82.

$$R_T = R_1 + R_2 = 4 + 6 = 10\ \Omega$$

Since the current flow is the *same* throughout the circuit we use Ohm's Law to find the current.

$$I\ (amperes) = \frac{E\ (volts)}{Total\ resistance\ (ohms)}$$

$$I\ (amperes) = \tfrac{60}{10} = \underline{6\ A}$$

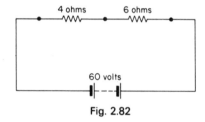

Fig. 2.82

When current flows through a resistor a voltage-crop occurs which can be found by using Ohm's Law. Thus,

$$E\ (volts) = I\ (amperes) \times R\ (ohms)$$

If the resistors are joined in series the sum of the voltage-drops across each resistor is equal to the supply voltage.

Example. Find the voltage-drop across each resistor in the circuit shown in Fig. 2.83

$$R_T = 2 + 4 + 6 = 12\ \Omega$$

$$I\ (amperes) = \frac{E\ (volts)}{R_T\ (ohms)} = \frac{24}{12} = 2\ A$$

Voltage-drop across R_1 = I (amperes) × R_1 (ohms)
= 2×6 = $\underline{12\ V}$

Voltage-drop across R_2 = I (amperes) × R_2 (ohms)
= 2×4 = $\underline{8\ V}$

Voltage-drop across R_3 = I (amperes) × R_3 (ohms)
= 2×2 = $\underline{4\ V}$

The total voltage-drop across the three resistors is 12+8+4 = 24 V which equals the supply voltage from the battery.

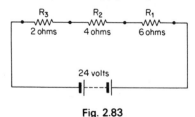

Fig. 2.83

Resistances in parallel. In Fig. 2.84 three resistances are shown connected across the same source of supply. Resistors connected in this way are said to be connected in *parallel*. The current divides into three paths I_1, I_2 and I_3. In general there will be as many paths as there are resistances in parallel.

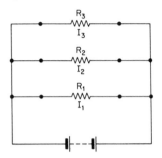

Fig. 2.84   Resistances in parallel

If one path is broken by a switch being opened (Fig. 2.85), a current will still pass through the remaining paths. Hence each path forms its own separate circuit.

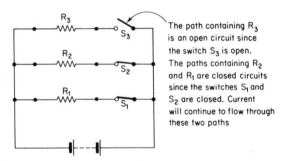

The path containing R_3 is an open circuit since the switch S_3 is open. The paths containing R_2 and R_1 are closed circuits since the switches S_1 and S_2 are closed. Current will continue to flow through these two paths

Fig. 2.85   Open and closed paths for resistances which are in parallel

When resistances are placed in parallel the total resistance is given by:

$$\frac{1}{R_T} = \frac{1}{R_1} + \frac{1}{R_2} + \frac{1}{R_3} + \ldots$$

Example. If in Fig. 2.84 $R_1 = 2\Omega$, $R_2 = 3\Omega$ and $R_3 = 4\Omega$, what is the total resistance of the circuit?

$$\frac{1}{R_T} = \frac{1}{2} + \frac{1}{3} + \frac{1}{4} = \frac{6+4+3}{12} = \frac{13}{12}$$

$$R_T = \frac{12}{13}\ \Omega$$

In a parallel circuit the current flowing out of a battery (or other supply) must be equal to the current flowing back to the battery. This current may be found by using Ohm's Law:

$$I\ (amperes) = \frac{E\ (volts)}{R_T\ (ohms)}$$

Example. If in Fig. 2.86, $R_1 = 2\Omega$ and $R_2 = 4\Omega$, find the current I.

$$\frac{1}{R_T} = \frac{1}{2} + \frac{1}{4} = \frac{2+1}{4} = \frac{3}{4}$$

$$\therefore R_T = \frac{4}{3}\Omega$$

$$I \text{ (amperes)} = \frac{E \text{ (volts)}}{R_T \text{ (ohms)}} = \frac{12}{\frac{4}{3}}$$

$$= 12 \times \tfrac{3}{4} = \underline{9 \text{ A}}$$

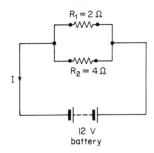

Fig. 2.86

In a parallel circuit the voltage-drop across each of the resistors is equal to the battery voltage. In Fig. 2.87 the voltage-drop across each of the resistors will be 6 V (the supply voltage).

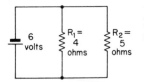

The voltage drop across the 4 ohm resistor is 6 volts. The voltage drop across the 5 ohm resistor is also 6 volts

Fig. 2.87   The voltage-drop across each of the resistors in a parallel circuit is equal to the supply voltage

Since we know the voltage drop across each resistor we can use Ohm's Law to find the current flowing in each path.

$$I \text{ (amperes)} = \frac{E \text{ (volts)}}{R \text{ (ohms)}}$$

Example. Find the current flowing in each of the two resistors shown in Fig. 2.87.

Current flowing in $R_1 = \frac{6}{4} = \underline{1 \cdot 5 \text{ A}}$
Current flowing in $R_2 = \frac{6}{5} = \underline{1 \cdot 2 \text{ A}}$

The total current flowing to and from the supply is 1·5 A+1·2 A = 2·7 A. That is, it is the sum of the currents in each of the paths.

Electrical measurements

1. *Measurement of current*. Current is measured by connecting an ammeter *into* the circuit as shown in Fig. 2.88.

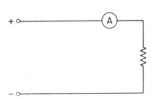

Fig. 2.88   An ammeter connected into the circuit to find the current flowing

Fig. 2.89 shows how ammeters may be used to find the current in various parts of a more complicated circuit.

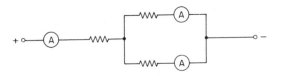

Fig. 2.89

2. *Measurement of voltage*. Voltage is measured by connecting a voltmeter *across* a circuit as shown in Fig. 2.90.

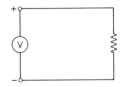

Fig. 2.90   A voltmeter connected across the circuit

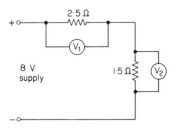

Fig. 2.91

In Fig. 2.91 the sum of the readings of the voltmeters V_1 and V_2 must be 8 V. The total resistance across the supply is 2·5+1·5 = 4Ω. The current taken from the supply is $\frac{8}{4}$ = 2 A. Hence V_1 reads 2×2·5 = 5 V and V_2 reads 2×1·5 = 3 V.

The three effects of an electric current

1. *Electromagnetism*. When an electric current flows through a wire, the wire acts as a magnet. If the wire is made into a coil, the coil acts like a strong bar

magnet. If we wish to increase the magnetic effect still further we can

(1) increase the current flow through the coil;
(2) insert a soft iron core into the coil.

Some typical electromagnets are shown in Fig. 2.92.

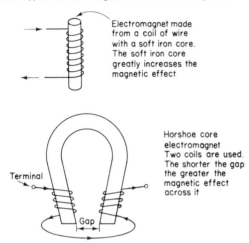

Electromagnet made from a coil of wire with a soft iron core. The soft iron core greatly increases the magnetic effect

Horshoe core electromagnet Two coils are used. The shorter the gap the greater the magnetic effect across it

Terminal

Gap

Fig. 2.92   Electromagnets

2. *Heating.* All conductors try to resist the flow of electricity and consequently when an electric current is passed through a conductor it gets hotter. If the current is high and the resistance is also high the conductor will get very hot. This is the principle of the electrical heating coil used in electric fires etc.

If the cross-sectional area of a wire is very small and a high current passes through it the wire becomes so hot that it melts. This is the principle used in the electric fuse. A circuit is thus protected against excessive current by means of fuses which melt when the current exceeds the safe level. A typical fuse is shown in Fig. 2.93.

Wire melts when current becomes too large

Fig. 2.93   A fuse

3. *Chemical effect.* When an electric current passes through a liquid conductor the liquid becomes chemically decomposed. When electricity is passed through a solution of metallic salt (e.g. copper sulphate) the current causes the metal to be deposited at the negative terminal. This effect is used in electroplating (see Chapter 2.7). The solution is called the *electrolyte* and the process is called *electrolysis*. In electroplating the article to be plated is attached to the negative terminal of the electrical supply and the metal to be used as the plating is deposited on the

article. The article thus becomes electroplated (Fig. 2.94).

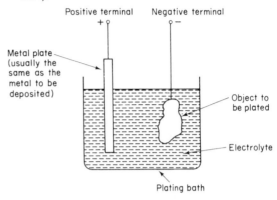

Fig. 2.94   Electroplating by electrolysis

Magnetism. Magnetism was perhaps first discovered by observing the behaviour of pieces of iron ore which are called 'lodestone'. These pieces of iron ore have the ability to set in a particular direction when freely suspended.

When a lodestone is put into a heap of iron filings, the filings move so that they are concentrated around two regions of the stone. These regions are called the poles of the lodestone. When the stone is suspended, a line drawn between the poles is found to point roughly north and south. The pole pointing towards the north is called the north-seeking pole (or simply the north pole). Similarly the pole pointing towards the south is called the south-seeking or south pole.

A group of materials known as *ferro-magnetic* materials all exhibit the same effect as the lodestone. These include iron, cobalt, nickel and certain alloys. Magnets made of these materials exhibit the effect to a much greater extent than does lodestone.

When a magnet is freely suspended it also points roughly north and south, the north pole of the magnet pointing north. If a second magnet is placed close to the suspended magnet, the direction in which the suspended magnet points is changed. A force is found to exist between the magnetic poles which causes like pole (i.e. both north or both south) to repel each other and unlike poles (one north and one south) to attract each other. The effect is rather like that of an electric charge mentioned on page 75. The above may be summarized as: *like poles repel each other and unlike poles attract each other.*

Magnetic fields. When a magnet is placed on a piece of card and iron filings are sprinkled over the card the filings set themselves in a definite pattern. This pattern is like a map which shows the conditions surrounding the magnet and is said to be due to the *field* of the magnet. The magnetic field has the property of exerting a force on the iron filings. It can also exert a force on another magnet. It follows also that the earth possesses its own magnetic field since it exerts a force on a suspended magnet to turn it to a north-south direction. Fig. 2.95 shows the lines of force around a bar magnet.

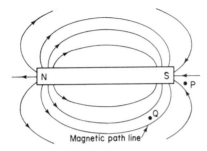

Fig. 2.95   Lines of force around a bar magnet

Magnetic flux. It is conventional to represent the direction of a magnetic field as away from a north pole and towards a south pole as shown in Fig. 2.95. It is, in fact, the direction in which an isolated north pole would move if it was free to do so.

A magnetic field contains complete magnetic paths and there are points in the field where the magnetic path lines are close together and others where the lines are far apart. Experience tells us that where the lines are far apart (which occurs at some distance from the magnet) the strength of the magnetic field is less. A strong magnet will produce magnetic path lines which are close together and a weak magnet will produce path lines which are farther apart. This gives the idea of magnetic flux. A strong magnet produces more magnetic flux than a weak one. Thus in Fig. 2.95 the field is stronger at P than it is at Q because the magnetic path lines are closer together. We say that the magnetic flux is denser at P than it is at Q and hence the field is stronger at P than it is at Q.

Permanent magnets. Soft iron does not retain its magnetism as steel does. Steel is frequently used as a permanent magnet but many other materials have been developed for this purpose. They include cobalt-steel, chromium steel, alaico and alcomax.

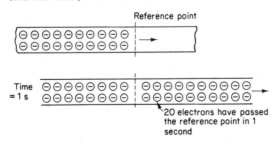

Fig. 2.96   D.C. current flow. The electrons flow in one direction only

Direct current (d.c.). The simplest form of direct current is that obtained from a cell where the electrons flow from the negative terminal to the positive terminal. *The current flows in one direction only* and the amount of current flowing is measured by counting the number of electrons which flow past a point in the circuit in 1 second (Fig. 2.96).

Alternating current (a.c.). *Alternating current flows back and forwards in a conductor.* Suppose that 10 electrons move in one direction past a point in $\frac{1}{2}$ second. When they reverse direction these 10 electrons will move past the same point in the next $\frac{1}{2}$ second. Thus 20 electrons have passed the point in 1 second and the current is exactly the same as if 20 electrons moving in one direction only had passed the point in 1 second (Fig. 2.97).

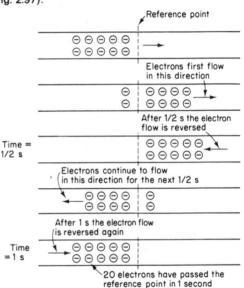

Fig. 2.97   A.C. current flow

Fig. 2.76 shows a simple a.c. generator which consists of a loop of wire rotating between two opposite magnetic poles. Each time one side of the loop passes from one pole to the other the current-flow in the loop reverses direction. When the loop passes two opposite poles the current flows first in one direction and then in the reverse direction, so giving a complete cycle of electron flow (Fig. 2.98). In the United Kingdom a.c. current reverses 100 times per second which is equivalent to 50 cycles per second (Fig. 2.99).

When an a.c. current reverses direction there is momentarily no current flowing in the circuit. When the current is used for electric lighting the lamp switches on and off 100 times per second. This is too fast for the human eye to detect and it therefore gets the impression that the lamp is permanently lit.

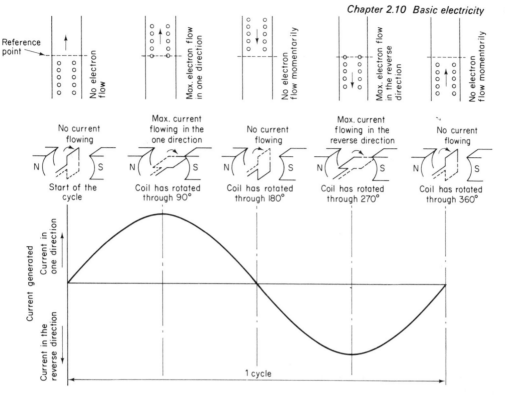

Fig. 2.98 The complete cycle for an a.c. generator

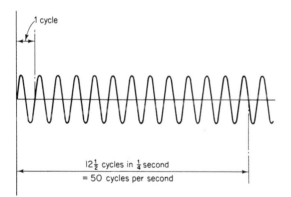

12½ cycles in ¼ second
= 50 cycles per second

Fig. 2.99 The frequency is the number of cycles per second. In the United Kingdom the standard frequency used is 50 cycles per second

Single- and three-phase supply. Supply cables to housing estates and factories are arranged as shown in Fig. 2.100. The three-phase wires are known as the red, yellow and blue phases. The fourth wire is the neutral wire.

As shown in Fig. 2.100, power is supplied from the neutral and any *one* of the phases. The supply authority tries to keep a balanced load on each of the three phases. When the load is correctly balanced the neutral wire carries little or no current. A third of the load is supplied by neutral and blue, a third by neutral and red and a third by neutral and yellow.

Heavy users of electricity use three-phase supply. Lathes, milling machines etc. using more than 400 W need three-phase supply. Lamps, soldering irons and domestic appliances such as refrigerators and washing machines use single-phase supply. The voltage between neutral and any phase is about 240 V whilst the voltage across any two phase wires is about 440 V.

Exercise 2.10
1. What is an electron?
2. Show in a diagram how the electrons of an atom orbit the nucleus.
3. What are loosely-bound electrons?
4. Sketch the arrangement of loosely-bound electrons when a flow of electricity occurs.
5. What is the practical unit of quantity of electricity?
6. Show in a diagram how chemical action produces a flow of electricity.
7. How does magnetism produce a flow of electricity?
8. What is a conductor and what is an insulator? Give three examples of materials which can be used as a

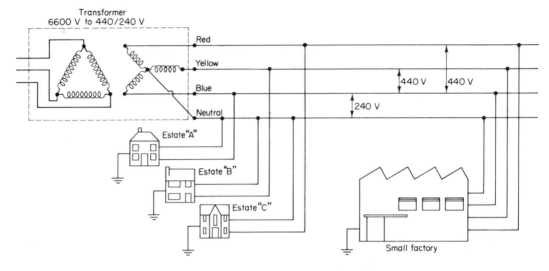

Fig. 2.100   A typical single-phase and three-phase
supply system

conductor and three which can be used as an
insulator.

9. A current of 5 A flows for 20 seconds. Find the
quantity of electricity used.

10. A current of 15 A flows for 5 seconds. What quantity
of electricity has been used?

11. A heating-element is designed to run at 240 V and
10 A. What power is needed?

12. A 3-kW electric fire is to run off a 250-V supply.
What current will be used?

13. A 1-kW electric fire is left on for 4 hours. How much
electrical energy (in kW h) has been used?

14. A 100-W lamp burns for 8 hours. How many units
of electricity have been used?

15. A consumer uses 400 units of electricity which is
charged for at 0·7p per unit. How much will the
consumer pay?

16. State the four factors which control the resistance
of a wire.

17. State Ohm's law.

18. A 500-Ω resistor uses a 250-V supply. What current
is needed?

19. A 240-V supply is wired to a 30-Ω resistor. What
current is flowing?

20. A current of 5 A is flowing through a resistance of
50 Ω. What voltage is needed?

21. A current of 8 A is supplied to a resistance of 20 Ω.
Find the voltage.

22. A 240-V supply causes a current of 8 A to flow
through a conductor. Find the resistance of the
conductor.

23. A car headlamp bulb takes a current of 3 A from a
12-V battery. What is the resistance of the bulb?

24. What is the total resistance of three resistors in
series if their resistances are 10 Ω, 20 Ω and 30 Ω
respectively.

25. Find the current flowing in the circuit shown in
Fig. 2.101.

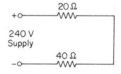

Fig. 2.101

26. Two resistors of 8 ohms and 4 ohms respectively are
connected in series to a 12-V supply. Find the current
flowing in the circuit.

27. Find the voltage-drop across each of the resistors
shown in Fig. 2.102.

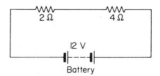

Fig. 2.102

28. Two resistors of 2 Ω and 3 Ω respectively are connected in parallel. Find their total resistance.

29. In Fig. 2.103 find the current I.

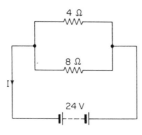

Fig. 2.103

30. Find the current flowing in each of the two resistors of Fig. 2.103.
31. What is the voltage-drop across each of the resistors in Fig. 2.103.
32. In Fig. 2.104 the circles represent either ammeters or voltmeters. For each circle state which instrument is represented.

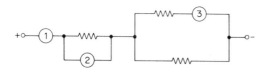

Fig. 2.104

33. In Fig. 2.105 state which circles represent ammeters and which represent voltmeters.

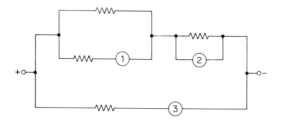

Fig. 2.105

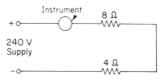

Fig. 2.106

34. Find the reading of the instrument shown in Fig. 2.106.
35. Find the readings of each of the instruments shown in Fig. 2.107.

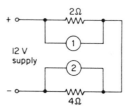

Fig. 2.107

36. Find the readings of each of the instruments shown in Fig. 2.108.

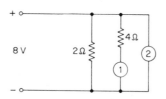

Fig. 2.108

37. State the three effects of an electric current.
38. What are electromagnets? How may the magnetic effect be increased?
39. How is electroplating done? Explain the process by means of a diagram.
40. What are ferro-magnetic materials? Name three such materials.
41. What is permanent magnet?
42. What is meant by magnetic flux? How is it used to measure the strength of a magnetic field?
43. What is the main difference between d.c. and a.c. currents?
44. What is meant by frequency?
45. Explain the connection between three-phase and single-phase supply.

Part 3   Engineering drawing

3.1   Geometrical constructions

To bisect a straight line
(1) Draw line AB.
(2) With centre A and at any radius greater than half AB draw arcs above and below AB.
(3) With centre at B draw arcs of same radius to cut previous arcs at C and D.
(4) Join C and D. The line CD bisects AB at E and is perpendicular to AB.

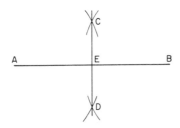

Fig. 3.1

To construct a perpendicular at the end of a line
(1) Draw line AB.
(2) With centre at A and at any radius draw the arc CD.
(3) With centre at C step off the same radius twice along the arc CD to give the points E and F.
(4) With centres E and F draw arcs of the same radius to cross at G.
(5) Join GA. This is the required perpendicular at the end A.

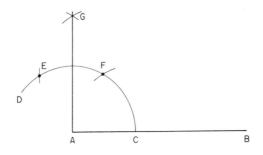

Fig. 3.2

**To construct a perpendicular at the end of a line.
Alternative method**
(1) Draw line AB four units long.
(2) With centre at A draw an arc three units long.
(3) With centre at B draw an arc five units long to cross previous arc at C.
(4) Join AC. This is the required perpendicular at the end A.
NOTE: This method is useful when space is too limited to use the previous method.

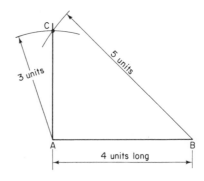

Fig. 3.3

To construct a 60° angle
(1) Draw line AB.
(2) With centre at A draw arc CD at any radius.
(3) With centre at C step off the same radius to cut arc at E.
(4) Join AE. The angle EAC is 60°.

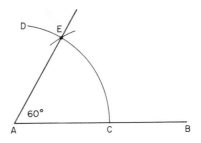

Fig. 3.4

To bisect a given angle
(1) Draw angle BAC.
(2) With centre at A and at any suitable radius draw an arc to cut AB at D and AC at E.
(3) With the centre at D and any radius draw an arc approximately central to angle BAC.
(4) With centre at E and at the same radius draw an arc to cut the previous arc at F.
(5) Join AF. This bisects angle BAC.

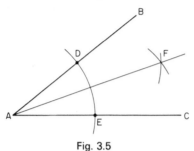

Fig. 3.5

To construct a 45° angle
(1) Draw line AB.
(2) Construct a perpendicular at A and at 90° to AB by either of the two methods indicated previously.
(3) Bisect the 90° angle as shown in Fig. 3.5.
(4) This produces the required 45° angle.

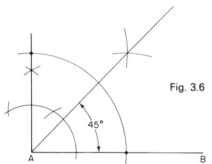

Fig. 3.6

To construct a 30° angle
(1) Construct a 60° angle, see Fig. 3.4.
(2) Bisect this angle as shown in Fig. 3.5.
(3) This produces the required 30° angle.
NOTE: By constructing 90° or 60° angles and bisecting them again and again, it is possible to produce many different angles. If necessary one angle may be built up upon another.

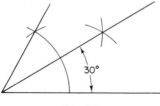

Fig. 3.7

Squares and rectangles
(1) Draw one side of the square AB.
(2) From A construct a perpendicular, see Fig. 3.2 or Fig. 3.3.
(3) With centre at A draw an arc from B to cut the perpendicular at C.
(4) With centres at B and C draw arcs crossing at D.
(5) Join CDB to complete the square. Rectangles may be constructed in a similar manner.

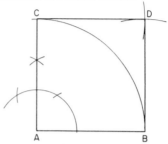

Fig. 3.8

Triangles
1. *To construct a triangle given the lengths of all three sides.*
   (1) Draw one side of the triangle AB.
   (2) With centre A and radius equal to the length of the second side draw an arc.
   (3) With centre B and radius equal to the length of the third side draw an arc to intersect the arc previously drawn. Let the point of intersection be C.
   (4) Join AC and BC. ABC is the required triangle.

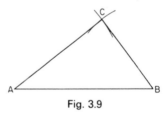

Fig. 3.9

2. *To construct a triangle given two sides and the included angle.* In Fig. 3.10 we are given the lengths of AB and AC and the angle at A.
   (1) Draw one side of the triangle AB.
   (2) Mark off the angle at A. Draw line AX.
   (3) Along AX measure the length of side AC.
   (4) Join BC. ABC is the required triangle.

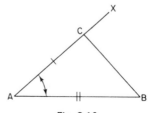

Fig. 3.10

3. *To construct a triangle given two sides and an angle which is not the included angle between the two sides.* In Fig. 3.11 we are given the sides AB and BC and the angle at A.
  (1) Draw one side of the triangle AB.
  (2) Mark off the angle at A and draw the line AX.
  (3) With centre B and radius equal to the length BC draw an arc to cut AX at C and C'.
  (4) Join BC and BC'.
  It will be seen from Fig. 3.11 that there are two triangles (ABC and ABC') which can be drawn from the given data. This case is known as the ambiguous case.

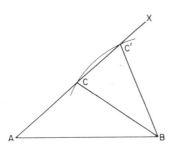

Fig. 3.11

4. *To construct a triangle given two angles and the length of one side.* In Fig. 3·12 we are given the angles at B and C and the length of the side AB.
  (1) Draw the side AB.
  (2) Mark off the angle at B and draw the line BX.
  (3) Calculate the angle at A. Thus:
     angle at A = 180°—(angle at B+angle at C).
  (4) Mark off the angle at A and draw AC to intersect with BX at C. The required triangle is ABC.

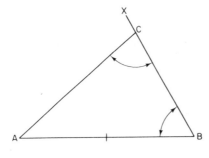

Fig. 3.12

Parallelograms

To draw the parallelogram ABCD given the lengths of the sides AD and DC and the angle at D. (Fig. 3.13)
  (1) Draw the side DC.
  (2) Mark off the angle at D and draw the line DX.

(3) From C draw the line CY parallel to the line DX.
(4) Along DX mark off the length DA.
(5) From A draw the line AB parallel to DC meeting the line CY at B. ABCD is the required parallelogram.

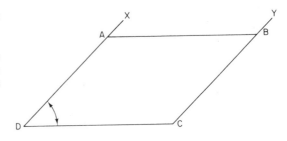

Fig. 3.13

Circles and tangents. A circle is a plane figure bounded by a curved line called a *circumference*, which is always the same distance from a fixed point called the *centre* of the circle. The distance from the centre to the circumference is the *radius* of the circle. An *arc* is the name for the part of a circumference between any two points. A *chord* is a straight line which joins any two points on a circumference. A *segment* of a circle is the area which is bounded by a chord and the arc it cuts off. A *sector* of a circle is the area which lies between two radii and the arc between them. A *quadrant* is the area bounded by two radii at right-angles to each other and the arc which lies between them. It is a quarter of a circle.

A *tangent* is a straight line which touches a circle but does not cut it. The point where it touches the circumference is called the point of tangency. At this point the tangent is at 90° to the radius.

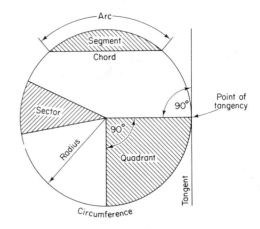

Fig. 3.14

To find the centre of a circle

(1) Draw any two chords AB and BC. NOTE: Greater accuracy is obtained if the chords are approximately at right-angles.
(2) Bisect each chord (see Fig. 3.1).
(3) Produce the bisectors to meet at O. O is the required centre of the circle.

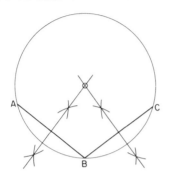

Fig. 3.15

To divide a straight line into a given number of equal parts

(1) Draw the given line AB.
(2) At any convenient angle to AB draw the line AC.
(3) Step off along AC the required number of divisions. These may be any convenient length but must be all equal. (7 divisions are shown)
(4) Join the last number to point B.
(5) Draw parallel lines from the remaining divisions to the given line AB. AB is now divided into the required number of equal parts.

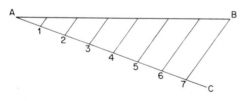

Fig. 3.16

Regular polygons

1. THE HEXAGON is a six-sided polygon (Fig. 3.17) with each of the six sides equal in length.
(a) *To construct a hexagon given the distance across corners.*
(1) Draw a circle whose diameter is equal to the distance across the corners of the hexagon.
(2) Mark any point on the circumference of the circle. Using the marked point as the first centre and keeping the compasses at the radius of the circle, mark off 6 points on the circumference. Join the points together to complete the hexagon (Fig. 3.18).

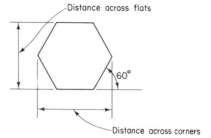

Fig. 3.17

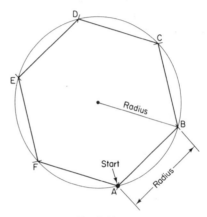

Fig. 3.18

(b) *To construct a hexagon using set squares.* This method is used when the distance across flats is given. The diagram (Fig. 3.19) shows how to construct the hexagon.

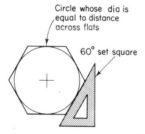

Fig. 3.19

2. THE OCTAGON is an eight-sided polygon (Fig. 3.20) with each of its eight sides equal in length. Octagons are easily constructed by using 45° set squares (Fig. 3.21).

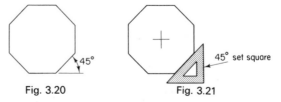

Fig. 3.20   Fig. 3.21

Exercise 3.1

1. Using rule and compasses only, bisect a line 10 cm long.
2. Erect a perpendicular at the end of line 8 cm long.
3. Construct (a) a 60° angle (b) a 45° angle.
4. Using a protractor draw an angle of 75°. Bisect this angle using rule and compasses only.
5. Using rule and compasses only draw an angle of 75°.
6. Construct a rectangle which is 10 cm long and 8 cm wide.
7. Construct a square of 15 cm side.
8. Draw the triangle whose sides are 6 cm, 8 cm and 9 cm long respectively.
9. Construct the triangles shown in Fig. 3.22.

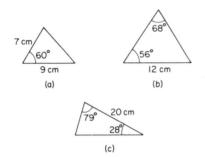

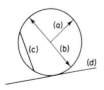

Fig. 3.22

10. Construct the parallelograms shown in Fig. 3.23.

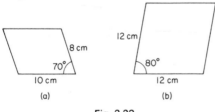

Fig. 3.23

11. Name the parts of the circle shown in Fig. 3.24.

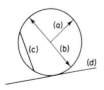

Fig. 3.24

Fig. 3.25

12. Name the parts of the circle shown in Fig. 3.25.

13. Divide the following straight lines:
   (a) 15 cm long into 5 equal parts;
   (b) 20 cm long into 8 equal parts;
   (c) 22 cm long into 7 equal parts.
14. Construct a hexagon which is 10 cm across corners.
15. Construct a hexagon which is 8 cm across flats.
16. Construct an octagon which is 12 cm across flats.
17. Construct an octagon which is 11 cm across corners.
18. Draw the template shown in Fig. 3.26.

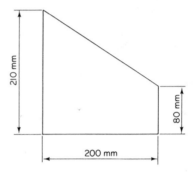

Fig. 3.26

19. Draw the plate detail shown in Fig. 3.27.

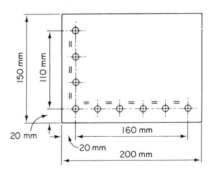

Fig. 3.27

20. Draw the template shown in Fig. 3.28.

Fig. 3.28

3.2 Orthographic projection

Engineering drawings are used to show all the details of an article to be manufactured.

The types of lines used in engineering drawing are shown in Fig. 3.29. Examples of their use are shown in Figs. 3.30 and 3.31.

Continuous (thick)	A	———————	Visible outlines
Continuous (thin)	B	———————	⎰Dimension lines ⎱Projection lines ⎰Hatching or sectioning
Short dashes (thin)	C	- - - - - - - -	Hidden details
Long chain (thin)	D	——— - ———	⎰Centre lines ⎱Path lines for indicating movement
Long chain (thick)	E	——— - ———	Cutting or viewing planes
Short chain (thin)	F	— - — - —	Developed views
Wavy (thick)	G	∼∼∼∼∼	Irregular boundary lines
	H	—∿—∿—	Long break lines

Fig. 3.29

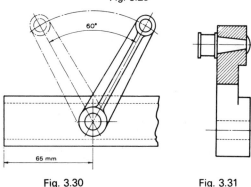

Fig. 3.30   Fig. 3.31

Orthographic projection. The object shown in Fig. 3.32 is very easy to understand, but it shows only three sides of the object out of a possible six. To overcome this disadvantage a system of drawing known as orthographic projection has been developed. In this system a full view of each side of the object is shown in turn. Normally three full views are sufficient to show all the details of the object.

There are two versions of orthographic projection, namely *first-angle* and *third-angle* projection. The only difference between the two is the relative positions in which the views are placed on paper. The main principle of orthographic projection—that the projection lines are always parallel to each other—is the same for both versions.

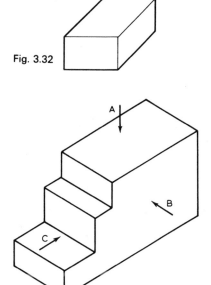

Fig. 3.32

Fig. 3.33

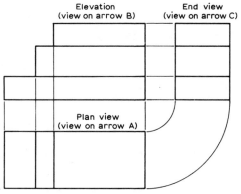

First angle projection

Fig. 3.34

Fig. 3.33 shows a pictorial drawing of a metal block. To draw this in *first-angle projection* a draughtsman looks at the front of the block in the direction shown by arrow B. The view he sees is known as the *elevation*.

Next he looks at the top of the block in the direction of the arrow A. The view he sees is called the *plan view*. As shown in Fig. 3.34 this is drawn directly under the elevation. Finally he looks in the direction of arrow C and the view he sees is called the *end view*. As shown (Fig. 3.34) this is drawn in line with the elevation.

To draw the block in *third-angle projection*, the draughtsman looks at it from the same directions as for first-angle projection (i.e. using the arrows A, B and C). Now, however, the plan view is drawn directly *above* the elevation and the end view is placed to the *left* of the

Plan view
(view on arrow A)

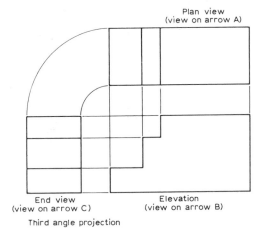

End view   Elevation
(view on arrow C)   (view on arrow B)

Third angle projection

Fig. 3.35

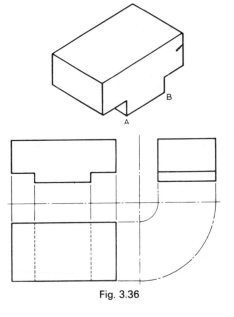

Fig. 3.36

elevation. Notice carefully that the three views are precisely the same for each type of projection. It is only the positions of the views which are different. This is shown in Fig. 3.35.

Projection lines are used when making a drawing to make sure that the views are correctly positioned. In Figs. 3.34 and 3.35 the faint projection lines are shown and it will be noticed that these allow the plan, elevation and end view to line up properly. On drawings issued to the workshops the projection lines are often omitted and in reading the drawings these have to be imagined.

Hidden details are shown by dotted lines, which help the reader to understand the drawing. In drawing the object shown in Fig. 3.36 we cannot see the face AB when looking at the top of the block. This face is shown by the dotted lines in plan view.

Section views. It is often found that the outside views of an object are inadequate for its clear description. Fig. 3.37 shows an open box. If this box is drawn orthographically (Fig. 3.38) the dotted lines show that the box has a bottom. On a complicated drawing this method

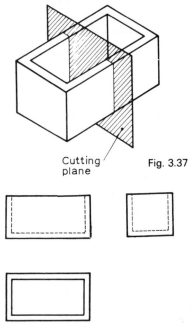

Cutting
plane   Fig. 3.37

Fig. 3.38

lacks clarity, the dotted lines tending to make the drawing confused. Also dimensioning becomes difficult. Cutting off part of the box with a cutting plane, as shown in Fig. 3.39, clearly reveals the internal shape when the orthographic views are drawn (Fig. 3.40). The faces that are imagined cut are then shown with the hatched or sectioned lines. The arrows AA show the direction in which the section is viewed. A typical sectioned view is shown in Fig. 3.41.

Half sections are used when the object is symmetrical about a centre-line. Both the internal and external details can then be shown in one view (see Fig. 3.42).

Parts not sectioned. The following parts are not sectioned when they lie in the plane of the section but they are shown in outside view: nuts, bolts, screws, washers, studs, pins, rivets (see Fig. 3.43); shafts, keys, cotters and spokes (see Fig. 3.44). Ribs and webs (see Fig. 3.45) which are cut parallel to the face of the rib or web are not sectioned. They are sectioned, however, if cut across the rib or web.

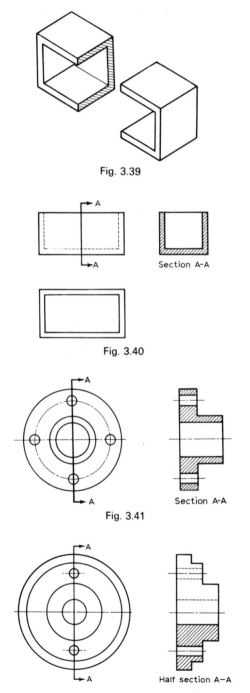

Fig. 3.39

Fig. 3.40

Fig. 3.41

Section A-A

Fig. 3.42

Half section A-A

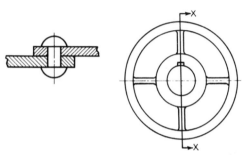

Fig. 3.43

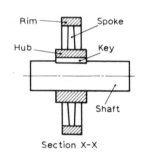

Section X-X

Fig. 3.44

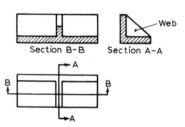

Section B-B   Section A-A

Fig. 3.45

Conventions. There are a number of accepted conventions in engineering drawing. A full list of these is shown in BS 308 (Engineering Drawing Practice). Fig. 3.46 shows the more important drawing conventions.

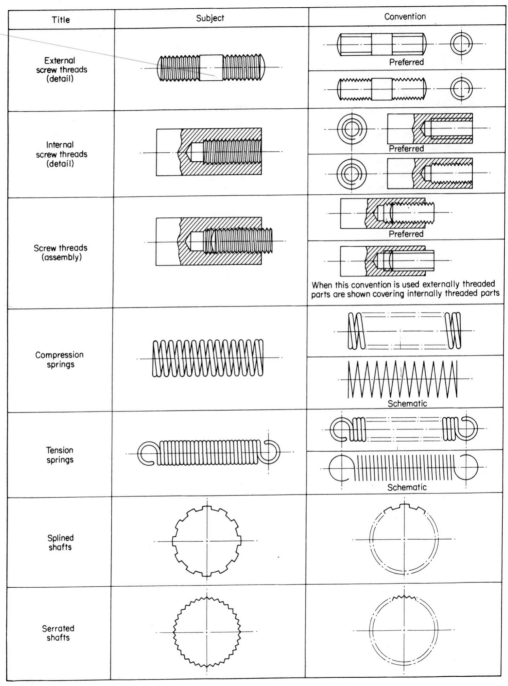

Title	Subject	Convention
External screw threads (detail)		Preferred
Internal screw threads (detail)		Preferred
Screw threads (assembly)		Preferred When this convention is used externally threaded parts are shown covering internally threaded parts
Compression springs		Schematic
Tension springs		Schematic
Splined shafts		
Serrated shafts		

Fig. 3.46 (a)   Standard drawing conventions
(reproduced from BS 308 with
permission)

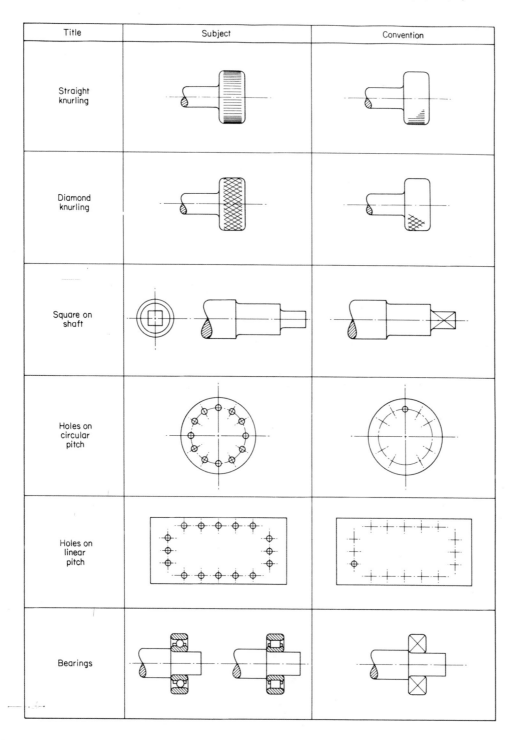

Title	Subject	Convention
Straight knurling		
Diamond knurling		
Square on shaft		
Holes on circular pitch		
Holes on linear pitch		
Bearings		

Fig. 3.46 (b)

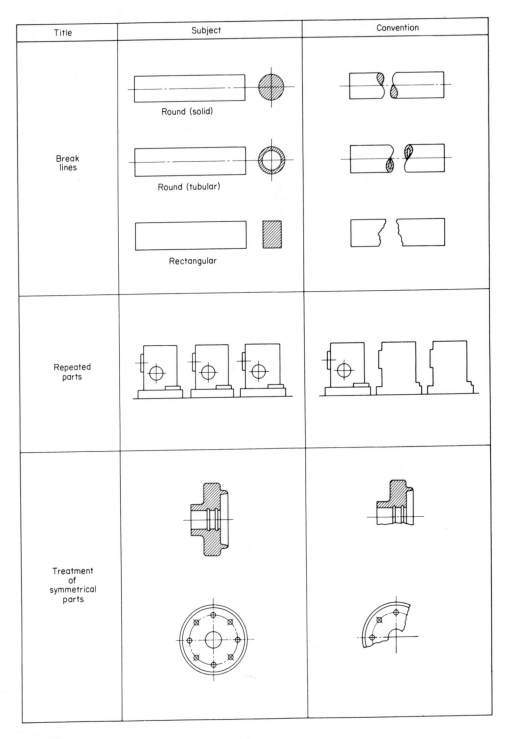

Fig. 3.46 (c)

Abbreviations shown on drawings. Abbreviations are frequently used on drawings. Those most commonly used are shown opposite and the way in which they are used is shown in Fig. 3.47:

Term	Abbreviation
Across flats	A/F
Centres	CRS
Centre line	CL or L
Cheese head	CH HD
Countersunk	CSK
Countersunk head	CSK HD
Counterbore	C'BORE
Diameter	DIA or Ø
Hexagon head	HEX HD
Machine	M/C
Not to scale	NTS
Outside diameter	O/D
Inside diameter	I/D
Pitch circle diameter	PCD
Radius	RAD or R
Round head	RD HD
Screwed	SCR
Spherical	SPH
Spotface	S'FACE
Undercut	U'CUT
Millimetre	MM

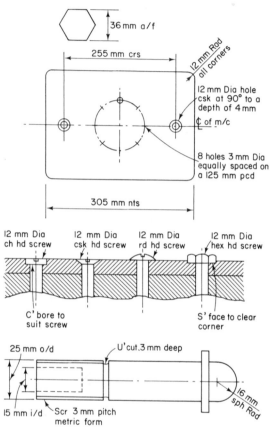

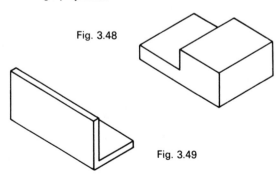

Fig. 3.47   Use of abbreviations on drawings

Exercise 3.2

1. Make three-view freehand drawings of each of the objects shown in Figs 3.48, 3.49, 3.50, 3.51, 3.52 and 3.53 using (a) first-angle projection (b) third-angle projection.

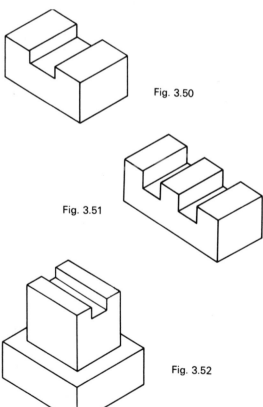

Fig. 3.50

Fig. 3.51

Fig. 3.48

Fig. 3.49

Fig. 3.52

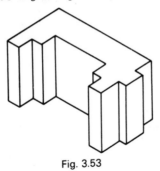

Fig. 3.53

2. Make three-view sketches of each of the articles shown in Figs 3.54, 3.55, 3.56, 3.57, 3.58 and 3.59 using (a) first-angle projection (b) third-angle projection.

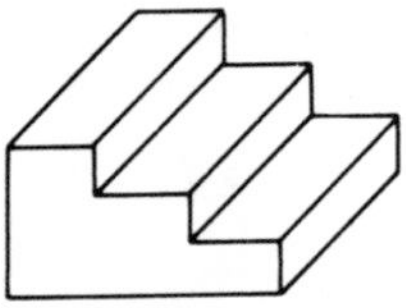

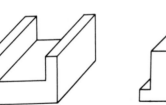

Fig. 3.54

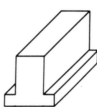

Fig. 3.55

Fig. 3.56

Fig. 3.57

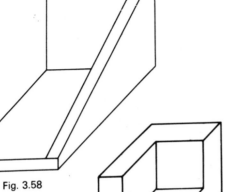

Fig. 3.58

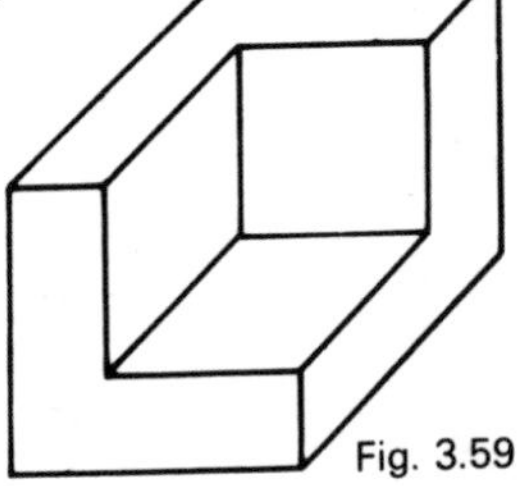

Fig. 3.59

3. Make three-view sketches of each of the articles shown in Figs 3.60, 3.61, 3.62, 3.63, 3.64 and 3.65. You may use either first- *or* third-angle projection but you must state the kind of projection used.

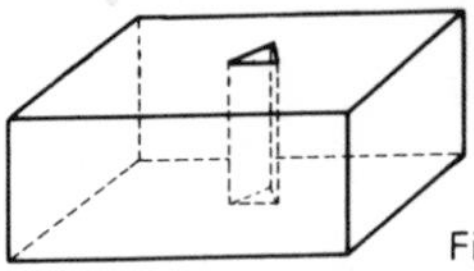

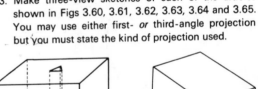

Fig. 3.60

Fig. 3.61

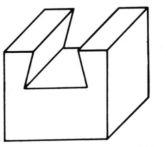

Fig. 3.62

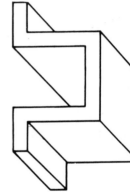

Fig. 3.63

Hole right thro'

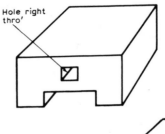

Fig. 3.64

Fig. 3.65

4. Make three-view drawings (a) in first-angle projection (b) in third-angle projection of each of the objects shown in Figs 3.66 and 3.67.

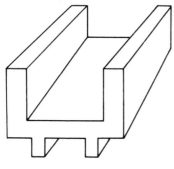

Fig. 3.66

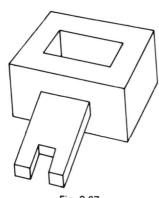

Fig. 3.67

5. Make a drawing in either first- *or* third-angle projection of the object shown in Fig. 3.68.

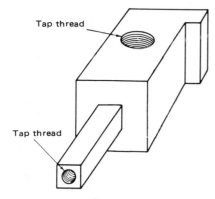

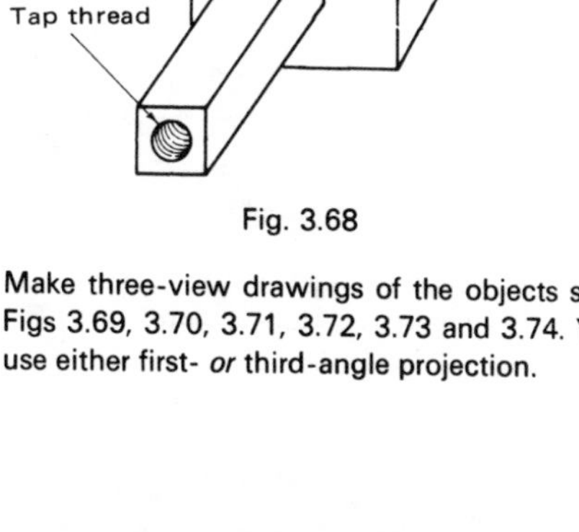

Tap thread

Tap thread

Fig. 3.68

6. Make three-view drawings of the objects shown in Figs 3.69, 3.70, 3.71, 3.72, 3.73 and 3.74. You may use either first- *or* third-angle projection.

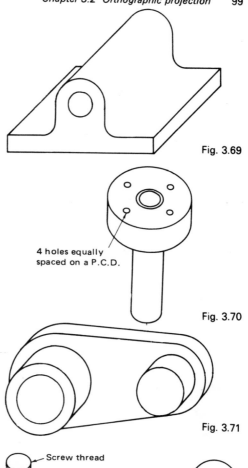

Fig. 3.69

4 holes equally
spaced on a P.C.D.

Fig. 3.70

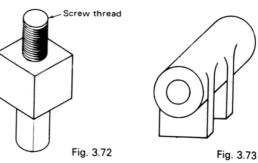

Fig. 3.71

Screw thread

Fig. 3.72

Fig. 3.73

Fig. 3.74

7. Make a three-view drawing of the spindle support bracket shown in Fig. 3.75. It must be drawn in third-angle projection.

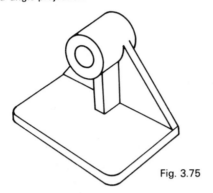

Fig. 3.75

8. Make a three-view drawing in first-angle projection or the bracket shown in Fig. 3.76

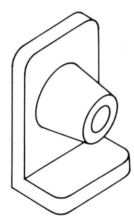

Fig. 3.76

9. Draw sectional A-A views as indicated for the parts shown in Figs 3.77 and 3.78.

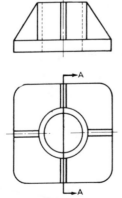

Fig. 3.77

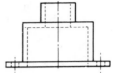

6 holes equally spaced on a P.C.D.

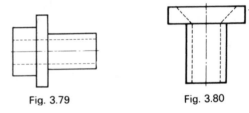

Fig. 3.78

10. Draw in half-section the parts shown in Figs 3.79 and 3.80.

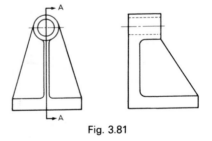

Fig. 3.79   Fig. 3.80

11. Sketch sectioned views as indicated in Figs 3.81 and 3.82.

Fig. 3.81

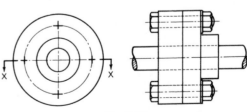

Fig. 3.82

12. Using the standard conventions make two- or three-view drawings of the parts shown in Figs 3.83 and 3.84.

(a)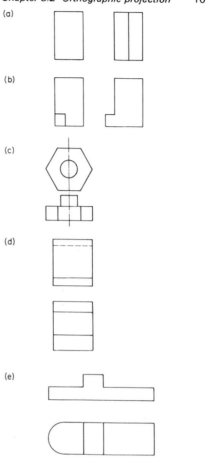

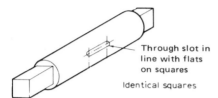

Through slot in
line with flats
on squares

Identical squares

Fig. 3.83

(b)

(c)

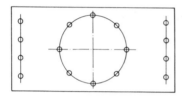

Fig. 3.84

(d)

13. Fig. 3.85 shows the side elevation of a solid component. State the cross-section of A and B. What is
represented by C? [*East Midlands Educational
Union*]

(e)

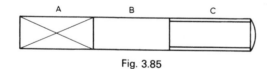

Fig. 3.85

Fig. 3.86

14. Write down the meanings of the following abbreviations:
A/F, PCD, RD HD, CL, CSK, M/C, NTS and
S'FACE.
15. The drawings of Fig. 3.86 each show two views of
an object. Sketch the third view. Notice that some
of the drawings are in first-angle projection whilst
some are in third-angle projection.
16. Fig. 3.87 shows three-view drawings of various
objects. Some of the drawings are in correct projection, others are not. For each drawing state
whether the projection is correct or not. For those
that are correct state the type of projection used
(i.e. first-angle or third-angle).
17. For each of the drawings in Fig. 3.88 sketch the
sectional view asked for. Some of the drawings are
drawn in first-angle projection and some are in third-
angle projection.

(a)

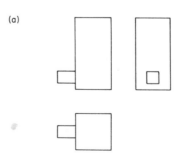

Fig. 3.87 (*continued overleaf*)

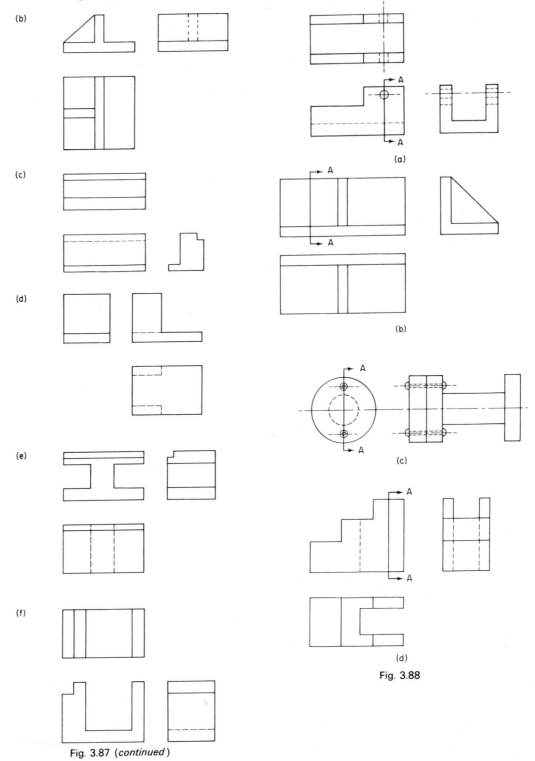

(b)

(c)

(d)

(e)

(f)

Fig. 3.87 *(continued)*

(a)

(b)

(c)

(d)

Fig. 3.88

3.3 Dimensioning of engineering drawings

An engineering drawing indicates the *shape* of an object. Dimensions indicate its size. Dimensions on mechanical-engineering drawings are usually expressed in millimetres no matter how large the object is. It is still accepted practice in the United Kingdom to use the centrally-placed decimal point with metric units, e.g. 24·88 mm. When a dimension is less than 1 mm the decimal sign is preceded by '0', e.g. 0·45 mm. For very large dimensions the numbers should be divided by a space between every third digit counting from the decimal marker, e.g. 18 357·5 or 9 001·52. However, where only four figures appear after or before the decimal point the space should be closed up, e.g. 3875 or 0·4113.

Elementary principles of dimensioning

(1) Every dimension which is necessary to define the finished workpiece should be shown on the drawing. It should appear once only (Fig. 3.89).

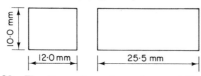

Fig. 3.89   The three dimensions shown define completely the size of the finished object

(2) If the drawing is properly made it should not be necessary for any dimension to be calculated from any other dimensions (Fig. 3.90). A drawing must never be scaled in order to find a dimension.

(3) There should not be more dimensions than are needed to define the workpiece (Fig. 3.91).

Auxiliary dimensions. In Fig. 3.91 (a), the overall dimension has been shown and hence one of the intermediate dimensions is unnecessary and it should not be

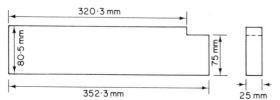

Fig. 3.90   Always work to the dimensions given. Do not attempt to calculate the sizes of the cut-out at the top right-hand corner of the plate

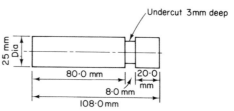

(a) Incorrect dimensioning: There are too many dimensions on the drawing

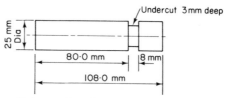

(b) Correct dimensioning: The dimensions stated completely define the size of the object

Fig. 3.91

shown. However where all the intermediate dimensions are shown it would be useful to know the overall dimension of the part. In such cases the overall dimension is called an auxiliary dimension and it is marked as shown in Fig. 3.92. It is stressed that the auxiliary dimension

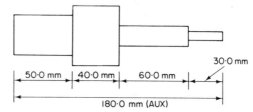

Fig. 3.92   The use of an auxiliary dimension to show the overall size of the object

should not be used in making the object. It may be used, for instance, in cutting the piece to approximate length before turning.

The importance of dimensions. A drawing properly dimensioned will indicate not only the size of the object but it will also show how the object is to be made. The craftsman has to interpret the drawing so as to make the object as the draughtsman intended it to be made.

Dimensions fall into two kinds:

(a) dimensions which give sizes.

(b) dimensions which locate (or position) features relative to each other.

The difference between the two is illustrated in Fig. 3.93.

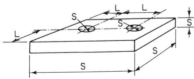

The dimensions marked S give the size of the block and the sizes of the holes. The dimensions marked L locate the positions of the holes

Fig. 3.93

Sometimes a single dimension will indicate size and location (Fig. 3.94).

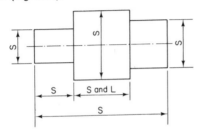

Fig. 3.94   A single dimension indicating both location and size

Datum faces. Fig. 3.95 shows the dimensioning of a plate detail. The length, width and thickness of the plate are *size* dimensions and so is the diameter of the hole. It will be noticed that the dimensions *locating* the position of the hole are given from two edges of the plate, i.e. the bottom edge and the left-hand edge. These edges are called *datum faces* since we must use these in positioning the hole.

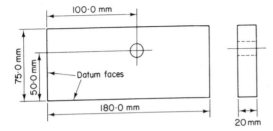

Fig. 3.95   Use of datum faces

Datum lines. When dimensions are used to position holes, the dimensions are frequently given from datum lines (Fig. 3.96). These datum lines must be used when marking out the hole positions.

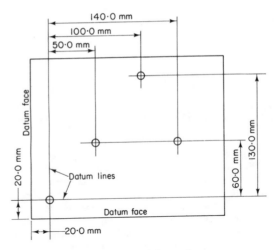

Fig. 3.96   Use of datum lines

Centre-lines. A centre-line is often used as one of the datum lines (see Fig. 3.97) and must be used when marking out the detail.

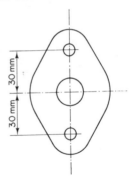

Fig. 3.97   Features positioned from centre-lines

Toleranced dimensions. It is impossible to manufacture components to an exact size. When we have to work to great accuracy the amount of error that we can tolerate is generally stated on the drawing. For instance, a turner may be given instructions on a drawing to turn a cylindrical bar to a diameter of 50mm $^{+0.10}_{-0.05}$. This means that the job will be satisfactory if the finished size lies between the following dimensions:

Greatest diameter = 50+0.10 = 50.10 mm

Least diameter   = 50−0.05 = 49.95 mm

The dimension 50.10 mm is called the *upper limit* and the dimension 49.95 is called the *lower limit*. The difference between the upper and lower limits is called the *tolerance*. In the above example the tolerance is 50.10−49.95 = 0.15 mm.

Example. A hole is to be made to 10.00±0.02 mm. Find the upper and lower limits and the tolerance.

The dimension 10·00±0·02 mm means

$$10.00 \begin{array}{l} +0.02 \text{ mm} \\ -0.02 \text{ mm} \end{array}$$

Upper limit = 10·00+0·02 = <u>10·02 mm</u>
Lower limit = 10·00−0·02 = <u>9·98 mm</u>
Tolerance = 10·02−9·98 = <u>0·04 mm</u>

There are several ways in which individual dimensions are toleranced.

(1) By stating the upper and lower limits (Fig. 3.98).

Upper limit = 78·89 mm
Lower limit = 78·84 mm
Tolerance = 0·05 mm

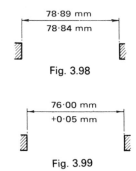

Fig. 3.98

Fig. 3.99

(2) By stating a limit of size with a tolerance in one direction only (Fig. 3.99).

Upper limit = 76·00+0·05 = 76·05 mm
Lower limit = 76·00 mm
Tolerance = 0·05 mm

(3) By stating a size with limits of tolerance above and below that size (Fig. 3.100).

Upper limit = 65·00+0·10 = 65·10 mm
Lower limit = 65·00−0·10 = 64·90 mm
Tolerance = 0·20 mm

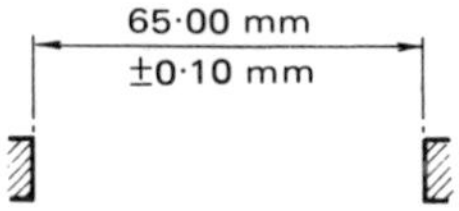

Fig. 3.100

Angular dimensions. Examples of angular dimensions are given below:

(1) $\frac{43°}{42°}$ The angle must lie between 42° and 43°.

(2) $\frac{28°}{-2°}$ The angle must lie between 26° and 28°.

(3) 57°±1° The angle must lie between 56° and 58°.

(4) $\frac{30° \; 30'}{30° \; 10'}$ The angle must lie between 30° 30' and 30° 10'.

Accumulative tolerances. Continuous dimensions as shown in Fig. 3.101 always tend to build up the inaccuracies. Thus, in Fig. 3.101 if the operator works to the upper limit for each dimension the overall length will be 115·00+0·22 = 115·22 mm.

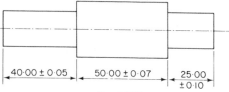

Fig. 3.101

The tolerance on the overall length may be too large. If the drawing is dimensioned as shown in Fig. 3.102 it is impossible for the overall length to have accumulative tolerances. At most, the overall length can be 115·15 mm.

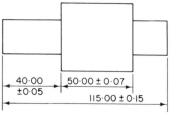

Fig. 3.102

In making the detail shown in Fig. 3.102 we can face both ends to give an overall length which lies between 114·85 and 115·15 mm. We can then face the shoulder to give a length of 40·00±0·05 mm and finally we can face the right-hand shoulder to give the length 50·00±0·07 mm.

Machining symbols. These are used to indicate that surfaces are to be machined. The general symbol used is shown in Fig. 3.103 and it should be placed on the same view as the dimensions which give the size or location of the surface concerned.

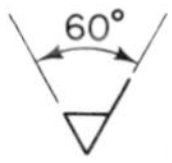

Fig. 3.103   Form of the machining symbol

Some examples of the way in which the machining symbol is used are shown in Fig. 3.104.

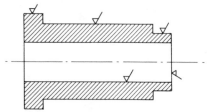

Fig. 3.104   Applications of the machining symbol

When all the surfaces are to be machined a general note as shown in Fig. 3.105 is often used.

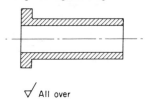

√ All over

Fig. 3.105

Scales. It is not always possible to make drawings the same size as the objects they represent. The drawing of a large casting must be drawn considerably smaller than the casting. On the other hand parts of small instruments, watches, etc. must be drawn larger than the parts themselves in order that all the detail can be shown. Both small and large objects must be drawn so that the drawings can be easily handled and read. Thus, when the dimensions of a drawing have to be enlarged or reduced we make use of *scales*.

The representative or scale fraction. Let us suppose that we wish to represent a length of 800 mm on an object by a line 160 mm long. Thus, the line is $\frac{160}{800} = \frac{1}{5}$ of the length on the object. We call this a scale of $\frac{1}{5}$ (usually written .1/5). We mean that each dimension on the drawing is reduced by one fifth. In general the scale is obtained by substituting in the following formula:

$$Scale = \frac{Length\ on\ the\ drawing}{Length\ of\ corresponding\ part\ on\ the\ object}$$

The scales recommended for use on drawings are 1/1 (full size), 1/2, 1/5 and 1/10. Where scales greater than full size are required, scales of 2/1, 5/1 and 10/1 are recommended.

Example. Calculate the length of line on the drawing to represent the following:

(a) A length of 946 mm on the object. Drawing scale is 1/2;

(b) A length of 1018 mm on the object. Drawing scale is 1/5;

(c) A length of 2438 mm on the object. Drawing scale is 1/10;

(d) A length of 15 mm on the object. Drawing scale is 2/1;

(e) A length of 3·8 mm on the object. Drawing scale is 10/1.

(a) Length of line on drawing = $946 \times \frac{1}{2}$ = $\underline{473\,mm}$
(b) Length of line on drawing = $1018 \times \frac{1}{5}$ = $\underline{203\cdot6\,mm}$
(c) Length of line on drawing = $2438 \times \frac{1}{10}$ = $\underline{243\cdot8\,mm}$
(d) Length of line on drawing = $15 \times \frac{2}{1}$ = $\underline{30\,mm}$
(e) Length of line on drawing = $3\cdot8 \times \frac{10}{1}$ = $\underline{38\,mm}$

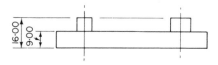

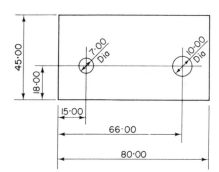

All dimensions in mm

Fig. 3.106

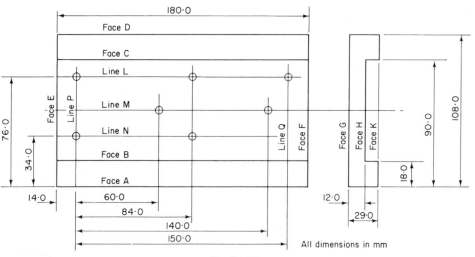

All dimensions in mm

Fig. 3.107

In the drawing office draughtsmen use a set of drawing scales made of boxwood or boxwood covered with celluloid. They look much like an ordinary rule except that they are graduated to give the scales required.

Exercise 3.3
1. In Fig. 3.106 take each dimension in turn and state whether it is a size dimension, a location dimension or both.
2. In Fig. 3.107 state which are (a) datum faces (b) datum lines (c) centre-lines.
3. In Fig. 3·108, state which is the auxiliary dimension.

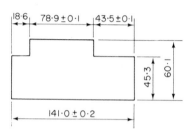

All dimensions in millimetres

Fig. 3.108

4. The drawing of a turned part is shown in Fig. 3.109. For each dimension calculate the lower limit, the upper limit and the tolerance.

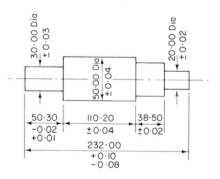

All dimensions in millimetres

Fig. 3.109

5. State how the part shown in Fig. 3.109 would be made in order to avoid accumulative tolerances.
6. Calculate the length of line on the drawing to represent the following:
   (a) A length of 220 mm on the object. Drawing scale is 1/2;
   (b) A length of 530 mm on the object. Drawing scale is 1/5;

(c) A length of 2017 mm on the object. Drawing scale is 1/10;
(d) A length of 20 mm on the object. Drawing scale is 2/1;
(e) A length of 3 mm on the object. Drawing scale is 10/1.
7. Make three-view dimensioned drawings of each of the objects shown in Fig. 3.110.

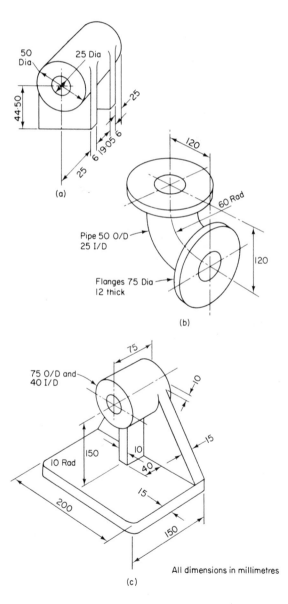

All dimensions in millimetres

Fig. 3.110

3.4 Auxiliary views

True length of a line. Fig. 3.111 shows an inclined line which has been drawn parallel to the vertical plane. As shown in Fig. 3.112 the elevation gives the true length of the line but the plan view shows a foreshortened length.

In Fig. 3.113 an inclined line has been drawn which is parallel to the horizontal plane. In this case the plan view of the line gives its true length but the elevation shows a foreshortened length (Fig. 3.114).

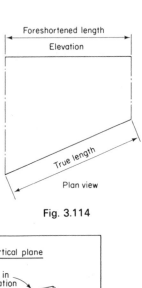

Fig. 3.114

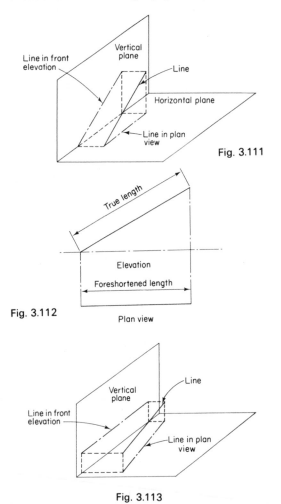

Fig. 3.111

Fig. 3.112

Fig. 3.113

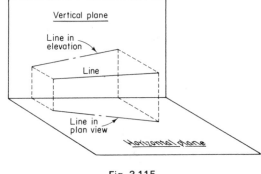

Fig. 3.115

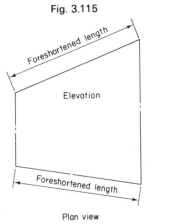

Fig. 3.116

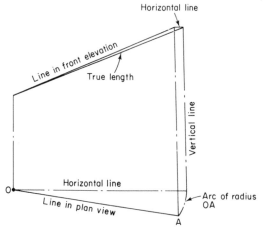

Fig. 3.117

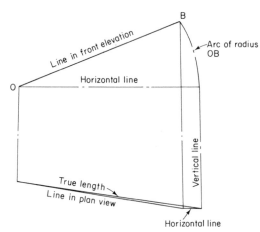

Fig. 3.118

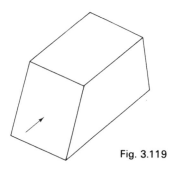

Fig. 3.119

Auxiliary views. Usually in engineering drawing we draw three views of an object projected on each of three planes at right-angles to each other. It is, however, sometimes useful to show a view of an object on an inclined plane.

Fig. 3.119 shows a block with an inclined face. We want to show a view of the block which will show the true shape of this inclined face. The method is shown in Fig. 3.120. The view obtained, which gives the true shape

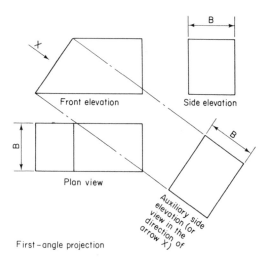

Fig. 3.120

of the inclined face is called an *auxiliary side elevation*. Sometimes on drawings this will be called 'view in the direction of arrow X'. It is important to note that the breadth B of the block is the same in both the side elevation and in the auxiliary side elevation.

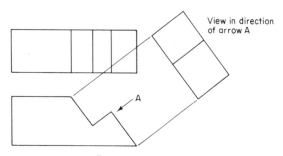

Third-angle projection

Fig. 3.121

Sometimes a line is parallel neither to the vertical plane nor the horizontal plane (Fig. 3.115). The front elevation and the plan view both show the line foreshortened in length (Fig. 3.116). The true length of the line is found by the construction shown in Fig. 3.117 *or* Fig. 3.118.

Fig. 3.121 shows an auxiliary view drawn in third-angle projection. Fig. 3.122 shows a practical application of auxiliary views.

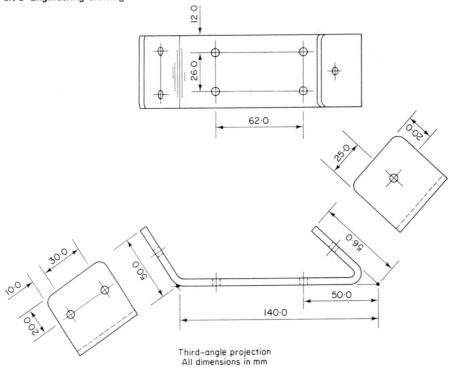

Third-angle projection
All dimensions in mm

Fig. 3.122   Dimensioned four-view drawing with
auxiliary views

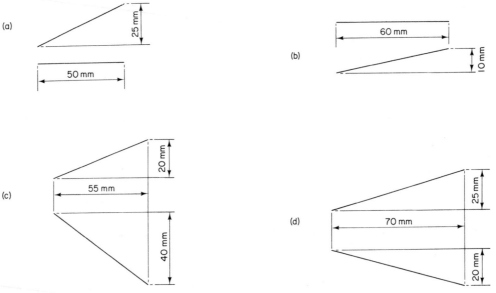

Fig. 3.123

Exercise 3.4
1. Find the true length of the lines shown in Fig. 3.123.
2. Make three-view drawings of each of the objects in Fig. 3.124: (a) in first-angle projection and (b) in third-angle projection. On each drawing make an auxiliary view in the direction of the arrow.
3. Draw true views in the direction of the arrows in Figs. 3.125, 3.126 and 3.127.

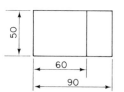

All dimensions in millimetres

Fig. 3.125

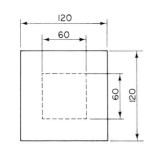

All dimensions in millimetres

Fig. 3.126

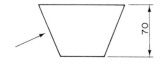

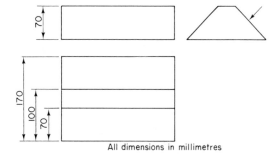

All dimensions in millimetres

Fig. 3.127

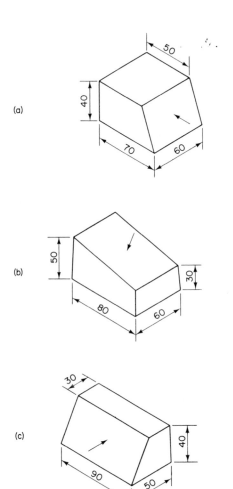

All dimensions in millimetres

Fig. 3.124

3.5 Pictorial drawing

Isometric projection. This is a way of representing an object by means of a pictorial drawing. It resembles a perspective sketch but it is often drawn by using drawing instruments. Isometric projection is used only to provide a pictorial view of an object. It is never used for a working drawing. It is of great value when sketches of machine parts are required and it is often used to make a pictorial view from two- and three-view drawings.

Isometric axes. In Fig. 3.128 three lines OX, OY and OZ have been drawn which make equal angles with each other. One of the lines is made vertical. These three lines are called the *isometric axes* and they form the basis for isometric drawing.

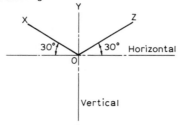

Fig. 3.128

Vertical edges of an object will be represented by vertical lines in the isometric drawing. When a rectangular object is being drawn the depth of the block is marked off on the vertical axis. The length and breadth are marked off on the other two axes (Fig. 3.129). For rectangular objects all the lines either lie along the three axes or are parallel with them. Fig. 3.130 shows two ways of drawing a rectangular block.

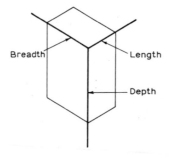

Fig. 3.129

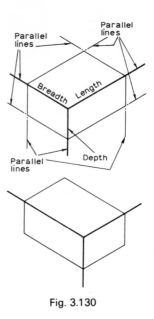

Fig. 3.130

Fig. 3.131 shows a three-view drawing of a metal block. To make an isometric drawing of this, first draw

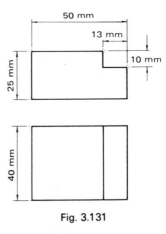

Fig. 3.131

the rectangular shape without the step. The step can then be drawn by the aid of the faint lines shown in Fig. 3.132. The last step in making the drawing is to line in the heavy lines and it is these which make the drawing clear.

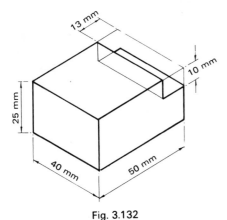

Fig. 3.132

Isometric representation of circles. Circles and parts of circles are usually drawn free-hand in isometric drawings. Most people find this difficult. Fig. 3.133 will give the clue necessary to overcome this difficulty. Fig. 3.133 (a) shows a true circle inscribed in a square. At the points marked X the circumference of the circle and the sides of the square appear to touch. This must be the same in the pictorial view shown in Fig. 3.133 (b). Fig.

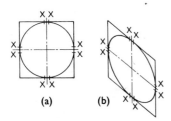

Fig. 3.133

3.134 shows a cube with a circle inscribed on each of its faces. Here again the circumference of the circles and the edges of the cube touch at the centre-lines of each of the faces.

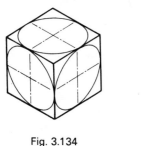

Fig. 3.134

Fig. 3.135

Fig. 3.135 shows the method of making an isometric drawing of a cylinder. The rectangular shape shown by

the faint lines is needed to obtain the circular shape of the ends of the cylinders.

Fig. 3.136 shows a three-view drawing of a machined block. The faint lines on the isometric drawing show how the circular shape is obtained.

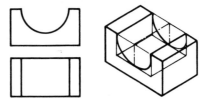

Fig. 3.136

Oblique projection. This is another way of representing objects pictorially. Fig. 3.137 shows the axes used. The axis AO is horizontal whilst axis BO is vertical. The axis

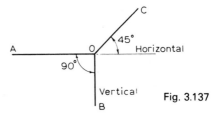

Fig. 3.137

OC is drawn at 45° to the axis AO produced. Measurements made along AO and BO (or parallel to them) are made full size. Those made along OC (or parallel to it) are generally made half size. Fig. 3.138 shows a 26 mm cube drawn in oblique projection. The lengths OA and OB are both made 26 mm but the length OC is made 13 mm.

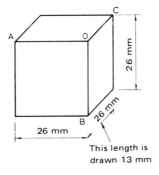

This length is drawn 13 mm

Fig. 3.138

Fig. 3.139 shows a three-view drawing of a metal block. To make the oblique projection first draw the three axes OA, OB and OC. *The front face of the block in the oblique drawing is the same as the front elevation.* Thus in the diagram the front face of the block is drawn using the axes OA and OB. The length OC is made 19 mm (that is half of 38 mm). The diagram shows how the rest of the block is completed.

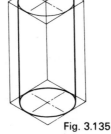

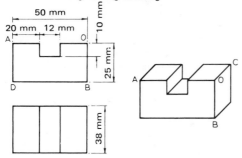

Fig. 3.139

When the circles have to be drawn it is often easier to make an oblique drawing than an isometric drawing. Figs 3.140 and 3.141 are examples. It will be noticed that the circles are drawn as true circles at the front of the drawings.

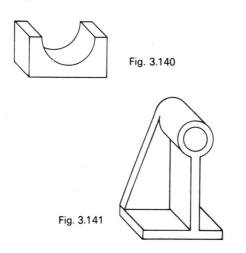

Fig. 3.140

Fig. 3.141

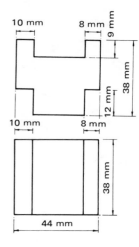

Fig. 3.143

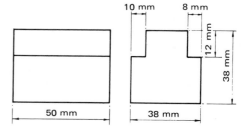

Fig. 3.144

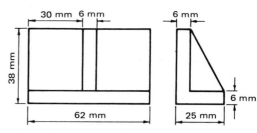

Fig. 3.145

Exercise 3.5

1. Make isometric drawings of the objects shown in Figs. 3.142, 3.143, 3.144 and 3.145.
2. Make oblique drawings of the objects shown in Figs. 3.142, 3.143, 3.144 and 3.145.

3. Make pictorial drawings of the articles shown in Figs. 3.146, 3.147 and 3.148: (a) in isometric projection; (b) in oblique projection.
4. Fig. 3.149 shows part of a planing machine toolholder. Draw a three-dimensional sketch (either isometric or oblique) of the toolholder showing as much detail as possible.
5. Make free-hand sketches of the following:
   (a) a soldering iron
   (b) a bench vice
   (c) a soft hammer
   (d) three types of cold chisel
   (e) a scribing block
   (f) a lathe faceplate
   (g) a pair of fullers
   (h) an external micrometer

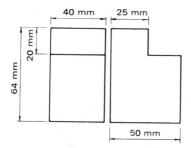

Fig. 3.142

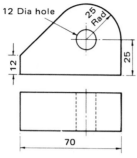

All dimensions in millimetres

Fig. 3.146

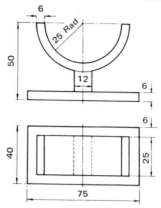

All dimensions in millimetres

Fig. 3.148

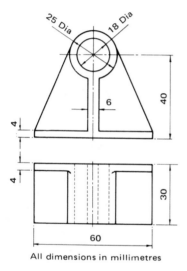

All dimensions in millimetres

Fig. 3.147

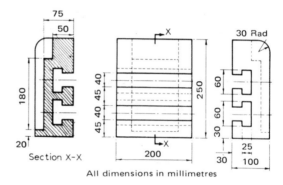

All dimensions in millimetres

Fig. 3.149

3.6 Welded joints

Welded joints. Fig. 3.150 shows some typical welded joints.

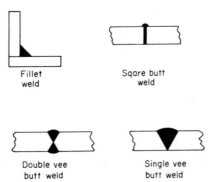

Fig. 3.150   Typical welded joints

Indicating welds on drawings. On a drawing it is necessary to state the position and type of weld required. This is done by:

(1) a weld symbol (see Fig. 3.151);
(2) an arrow and reference line (see Fig. 3.152).

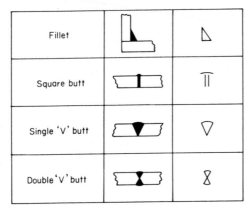

Fig. 3.151   Standard welding symbols

The weld symbols indicate the type of weld required (i.e. fillet, butt, etc.). Apart from stating the type of weld required it is also necessary to indicate the *position* of the weld. This is done by means of an arrow and a reference line (Fig. 3.152).

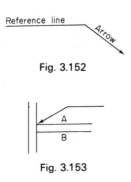

Fig. 3.152

Fig. 3.153

In Fig. 3.153:
    A represents the arrow side of the joint;
    B indicates the other side of the joint.

When the weld is required on the *arrow* side of the joint the welding symbol is placed *under* the reference line as shown in Fig. 3.154. If the weld is required on the *other* side of the joint the weld symbol is placed on the *top side* of the reference line (see Fig. 3.155).

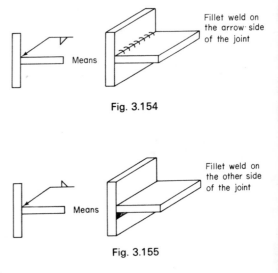

Fillet weld on the arrow side of the joint

Means

Fig. 3.154

Fillet weld on the other side of the joint

Means

Fig. 3.155

Now look at Fig. 3.156. It shows all the possible ways in which a single fillet weld can be indicated.

The way in which fillet welds on each side of the tee are indicated is shown in Fig. 3.157. Fig. 3.158 shows the way in which butt welds are indicated.

Sketch of weld   Symbolic representation

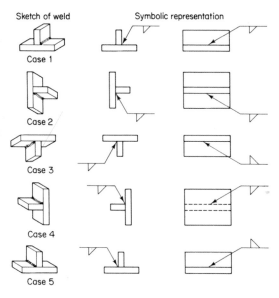

Case 1

Case 2

Case 3

Case 4

Case 5

Fig. 3.156   Typical applications of fillet welds (all drawn in third-angle projection)

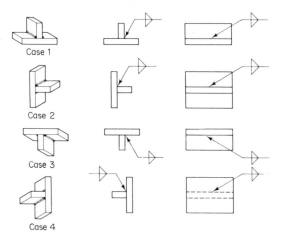

Case 1

Case 2

Case 3

Case 4

Fig. 3.157   Fillet welds each side of tee

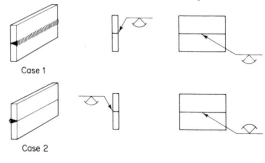

Case 1

Case 2

Fig. 3.158   Single 'V' butt welds (without sealing run)

Exercise 3.6

Fig. 3.159 shows the standard representation for various welded joints. Show in pictorial sketches the position of the weld.

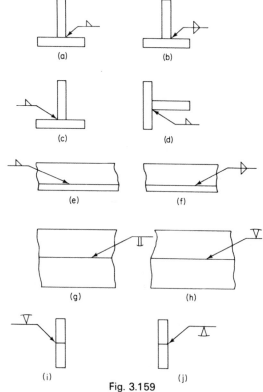

(a)   (b)

(c)   (d)

(e)   (f)

(g)   (h)

(i)   (j)

Fig. 3.159

3.7 Pattern developments

Sheet-metal developments. Before making an article in sheet metal we need a pattern, which is the shape of the flat metal needed to make the article. Fig. 3.160 (a) shows an angle bracket made from sheet metal. If we bend up face X until it is in the same plane as face Y, Fig. 3.160 (b), then we have the shape of the flat metal needed to make the bracket. This shape which is drawn and dimensioned in Fig. 3.160 (c) is called the pattern or the development of the bracket.

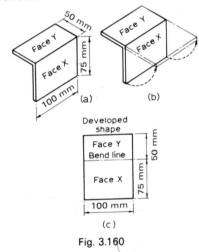

Fig. 3.160

Generally, the problem is to develop from a two- or three-view drawing. Fig. 3.161 shows a Z-shaped bracket. The faces X and Y are imagined to be straightened until they are in the same plane as face W. The development is best constructed on the elevation as shown.

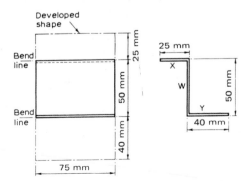

Fig. 3.162 (a) shows an open box made from tinplate. Draw the shape of a single sheet cut ready for bending, Fig. 3.162 (b).

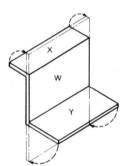

Fig. 3.161

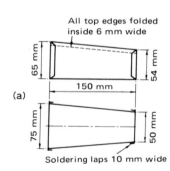

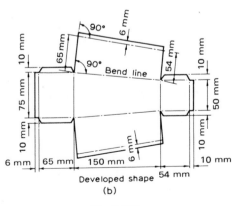

Fig. 3.162

Development of curved shapes. Curved shapes are often encountered. The following examples show how curved surfaces may be developed.

Fig. 3.163 (a) shows a cylindrical tin without a top or bottom. Find the developed shape (Fig. 3.163 (b)).

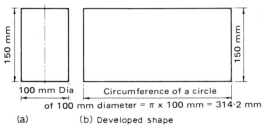

(a)   (b) Developed shape

Fig. 3.163

To develop the cone shown in Fig. 3.164 (a).

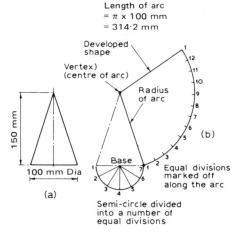

Fig. 3.164

To develop the cone (Fig. 3.164 (b)):
(1) Draw the front elevation of the cone.
(2) On the base of the cone draw a semicircle and divide it up into an equal number of divisions. The more divisions taken the more accurate the development.
(3) From the vertex of the cone draw an arc as shown.
(4) Step off along this arc the divisions taken on the semicircle and number as shown. Join the first and last divions to the vertex.

Development of cubes and pyramids

To develop the rectangular pyramid shown in Fig. 3.166 (a). The orthographic projection of the pyramid in first-angle is shown in Fig. 3.166 (b). From this projection we find the true length of the edge AB. Using this true length as radius we construct the developed shape as shown in Fig. 3.166 (c). When developing any kind of pyramid the radius used in the development must always be equal to the true length of an inclined edge of the pyramid.

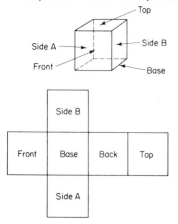

Fig. 3.165   Development of a cube

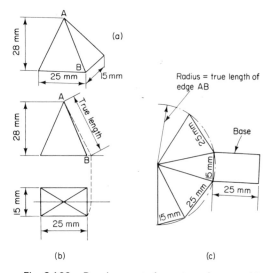

(b)   (c)

Fig. 3.166   Development of a rectangular pyramid

Exercise 3.7
1. Draw the shape of the pattern needed to make the objects shown in Fig. 3.167.
2. Draw the developments of the following pyramids:
   (a) Square base 30 mm side; height 20 mm.
   (b) Rectangular base 40 mm × 30 mm; height 25 mm.
   (c) Rectangular base 20 mm × 50 mm; height 40 mm.
3. Draw the developments of the following:
   (a) A cube of side 25 mm.
   (b) A rectangular box of length 40 mm, width 30 mm and height 20 mm.
   (c) A rectangular box of length 25 mm, width 35 mm and height 30 mm.
4. Draw the developed shape of a cylinder which has a diameter of 40 mm and a length of 60 mm.

5. Draw the developed shape of a cone which has a diameter of 80 mm and a height of 120 mm.

6. Draw the developed shape of the objects shown in Fig. 3.168.

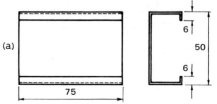

(a)

Channel bracket made from 1 mm thick steel plate

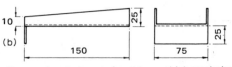

(b)

Support bracket made from 1 mm thick steel plate

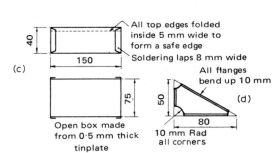

(c)

All top edges folded inside 5 mm wide to form a safe edge

Soldering laps 8 mm wide

All flanges bend up 10 mm

(d)

Open box made from 0·5 mm thick tinplate

10 mm Rad all corners

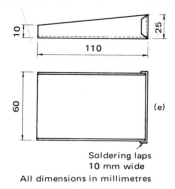

(e)

Soldering laps 10 mm wide

All dimensions in millimetres

Fig. 3.167

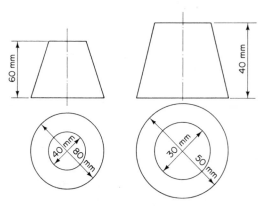

Fig. 3.168

3.8 Electrical circuit diagrams

For single-phase circuits there must be a live conductor and a neutral conductor. When drawing circuit diagrams the live conductor is often shown by a *continuous* line whilst the neutral conductor is shown by a *dotted* line.

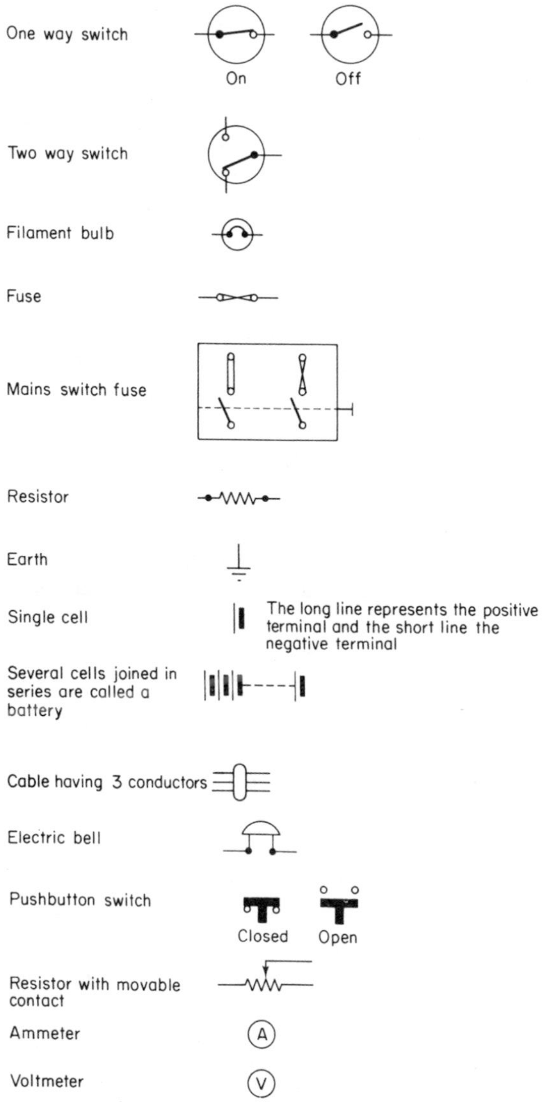

In the interests of simplicity the various items of a circuit are shown by means of standard symbols—details of which are shown in Fig. 3.169.

Examples of electrical circuit diagrams are shown in Figs. 3.170, 3.171 and 3.172. Notice that switches and fuses are always wired to the live conductor. It is dangerous to wire switches to the neutral conductor since current may still flow even when the switch is in the 'off' position.

One way switch   On   Off

Two way switch

Filament bulb

Fuse

Mains switch fuse

Resistor

Earth

Single cell   The long line represents the positive terminal and the short line the negative terminal

Several cells joined in series are called a battery

Cable having 3 conductors

Electric bell

Pushbutton switch   Closed   Open

Resistor with movable contact

Ammeter   (A)

Voltmeter   (V)

Fig. 3.169   Electrical symbols

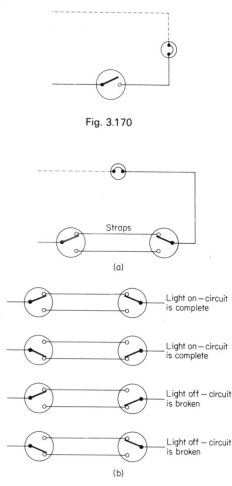

Fig. 3.170

Straps

(a)

Light on — circuit is complete

Light on — circuit is complete

Light off — circuit is broken

Light off — circuit is broken

(b)

Fig. 3.171 (b)

1. Simple lighting circuit which consists of one lamp controlled by a one-way switch. (Fig. 3.170)

2. A lighting circuit which consists of one lamp controlled by a two-way switch. (Fig. 3.171 (a)

The way in which the two-way switches operate is shown in Fig. 3.171 (b).

3. Circuit consisting of two lamps and an electric heater each controlled by its own switch. (Fig. 3.172)

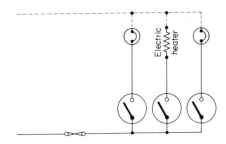

Fig. 3.172

4. A domestic lighting circuit. (Fig. 3.173)

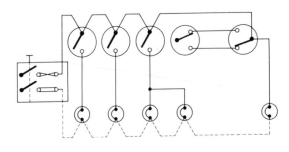

Fig. 3.173

Exercise 3.8

1. Fig. 3.174 shows a simple electrical circuit. Name the parts A, B and C.

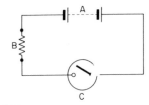

Fig. 3.174

2. In Fig. 3.175 name the parts lettered a, b, c, d and e. Redraw the circuit taking care to show the live conductor by a full line and the neutral conductor by a dotted line.

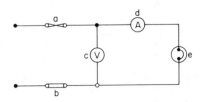

Fig. 3.175

3. Make diagrams for the following electrical symbols: (a) a fuse; (b) an earth; (c) a single cell; (d) a battery.

4. Draw a circuit diagram for a single electric light, mains operated, which incorporates a two-way switch.

5. Show circuit diagrams for: (a) two lamps in series controlled by one switch; (b) two lamps in parallel each controlled by separate switches.

6. Fig. 3.176 shows a one-bell push operating one bell. Name the parts lettered a, b and c.

Fig. 3.176

7. Redraw the circuit in Fig. 3.177 showing the live conductor with a full line and the neutral conductor by a dotted line.

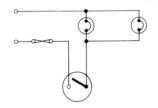

Fig. 3.177

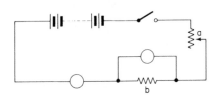

Fig. 3.178

8. In Fig. 3.178 the instruments represented by the circles are either ammeters or voltmeters. Redraw the circuit using the conventional representation for these instruments and name the parts a and b.
9. The instruments marked a, b, c and d in Fig. 3.179 are either voltmeters or ammeters. State which.
10. Draw pictorially:
   (a) A fuse holder;
   (b) A 13-ampere three-pin plug, stating the colours of the wires.

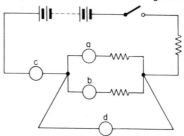

Fig. 3.179

Part 4   Craft theory

Safety

The following notes provide a background to safety-consciousness in the general workshop environment.

In the Chapters of Part 4 hints on safety are given for hand tools, machinery and welding, and other specific situations.

Clothing
(1) Loose clothing is dangerous because it can get caught in the moving parts of machines. For the workshop close-fitting overalls are the best form of clothing.
(2) The loose ends of scarves and ties can get caught in the revolving parts of machines. Make sure that these are well tucked in or, better still, remove them. Long shirt-sleeves are also dangerous.
(3) Long hair should be covered, particularly when working near revolving cutters.
(4) Rings and watches can get caught in machinery and should be removed.
(5) Shoes should have thick soles and reinforced, steel toecaps. Footwear with steel studs must be avoided because they can cause the wearer to slip and fall across a machine.
(6) Goggles should always be worn when chipping and grinding.
(7) When handling raw materials gloves should be worn as a protection against being cut by sharp edges which are nearly always present on sheet metal, castings, forgings, etc.

Keeping the workshop clean
(1) Keep all gangways clear and never place an object where somebody might fall over it.
(2) A slippery floor is dangerous so always keep the floor free from grease and oil.
(3) Watch out for bars of metal, etc. which protrude from vices, machines and racks.

Behaviour in the workshop. Fooling about in the workshop can cause you or a workmate serious injury. Never distract anyone who is working with tools or machinery or an accident may result.

Movement of materials
(1) Do not persist in attempting to lift a load which causes a feeling of strain.
(2) Before starting to move anything, remove any objects likely to cause obstruction and clean up slippery parts of the floor.
(3) Properly organize the work. Loads from lorries, etc., should be placed on platforms at about waist height.

They can then be lifted off the platform with little effort.
(4) Stacking at too high a level can cause considerable fatigue. If the stack becomes so high that load has to be lifted over head level a platform should be used to avoid this.
(5) Never risk a strain by trying to lift heavy or awkwardly-shaped objects. Always use the correct type of lifting tackle.
(6) Cranes, winches, slings, hooks, ropes and shackles must be used in the correct manner.
(7) Never stand under a load supported by lifting tackle.

Machine guards. These are placed on machines for your safety. Never remove the guard under any circumstances even if it means that you can work faster or see better. Special fencing or guards are used for such tools as a power press.

Health hazards. Some materials such as lead, dust and arsenic are harmful if ingested. Other materials such as corrosive acid are harmful to the eyes and skin. Some machine-cutting oils and epoxy resins can cause dermatitis. Substances such as trichlorethylene fumes are harmful if inhaled.
(1) Proper ventilation should be provided where this is necessary as in the case of harmful dusts.
(2) Personal precautions should be taken. When working with dangerous substances always use the washing facilities before eating.
(3) Barrier-creams rubbed on the hands can prevent dermatitis by preventing dirt and germs from entering the pores of the skin.

Fire-fighting equipment. Most firms have properly trained fire service teams. These teams can usually control a fire until the local fire service arrives. Every factory should have fire-drill regulations which you should obey in the case of fire. A fire needs oxygen for it to continue to burn. Hence, if the oxygen (from the air) can be removed or if the temperature can be reduced the fire will go out. Water reduces the temperature and the steam also helps to smother the flames.

For some burning substances water should not be used. Foam is used for oil and chemical fires. They act by preventing oxygen from the air reaching and feeding the flames. Carbon-dioxide extinguishers are used on burning vapours. The carbon dioxide is heavier than air and so it displaces the air and thus stops the fire. Dry-powder extinguishers are used for small fires caused by liquids and solids such as paper.

4.1 Measurement

Standards of length. Before anything can be measured a standard must be fixed. The standard of length in the SI system of units is the International Prototype Metres (Fig. 4.1) which is kept at Sèvres, near Paris. The standard metre is the distance between two fine lines engraved on one surface of the metal.

For most engineering purposes the metre is too large a unit and hence it is subdivided into millimetres such that:

$$1 \text{ metre} = 1000 \text{ millimetres}.$$

The millimetre may be subdivided into even smaller parts. We often need to measure to an accuracy of $\frac{1}{100}$ of a millimetre.

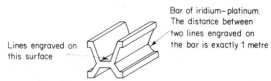

Fig. 4.1   The International Prototype Metre

Tools used in measuring. Some of the tools used in measuring are shown in Fig. 4.2. The *rule* is the craftsman's basic measuring tool. Measurements on a rule are made between two lines and, so far as the craftsman is concerned, the rule is the basic *line standard*. With a good rule measurements can be made to an accuracy of 0·2 mm but it depends very much on the skill of the user (Fig. 4.3).

Calipers may be used to an accuracy of 0·25 mm when used in conjunction with a rule.

Micrometers and vernier calipers are used when measurements accurate to $\frac{1}{100}$ of a millimetre are required. Calipers and micrometers measure between ends as opposed to lines.

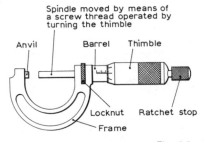

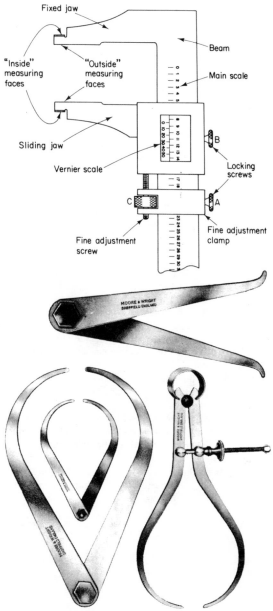

Fig. 4.2   The tools used in measuring

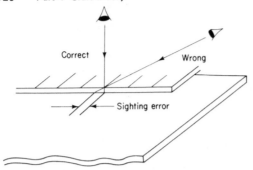

The eye should be vertically over the rule or sighting errors result

Fig. 4.3   How errors are introduced when reading a rule

The metric micrometer. The screw thread which moves the spindle has a pitch of 0·5 mm. That is, one turn of the thimble causes the spindle to move a distance of 0·5 mm. Fig. 4.4 shows how the metric micrometer scales are marked.

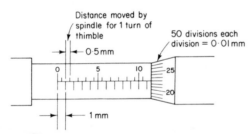

Fig. 4.4   The scales of a metric micrometer

To read the micrometer set as shown in Fig. 4.5:
(1) Read off the number of complete millimetres exposed (9 large divisions = 9 mm).
(2) Read off the number of half-millimetre divisions exposed beyond the last millimetre division (1 half division = 0·5 mm).
(3) Note the number of the thimble division that co-incides with the axial line on the sleeve. This number gives the number of hundredths of a millimetre (thimble reading is 16 = 0·16 mm).
(4) To obtain the micrometer reading add together the readings in (1), (2) and (3). The reading for this setting is 9+0·5+0·16 = 9·66 mm.

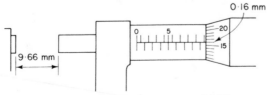

Fig. 4.5   A metric micrometer set to read 9·66 mm

The vernier caliper. Each division on the vernier scale is a little smaller than each division on the main scale. The difference between a main scale division and a vernier division gives the accuracy to which the instrument can be read.

There are various kinds of metric verniers. The one shown in Fig. 4.6 has main scale divisions spaced 0·5 mm apart. The vernier has 25 divisions equal in length to 24 main scale divisions. That is, 25 vernier divisions have a length of 12 mm. Each division on the vernier scale = $\frac{12}{25}$ = 0·48 mm. The difference between a main scale division and a vernier division = 0·5 − 0·48 = 0·02 mm. Hence the instrument can be read to an accuracy of 0·02 mm (or $\frac{1}{50}$ of a millimetre).

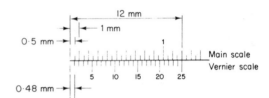

Fig. 4.6   The scales of a metric vernier caliper

To read the metric vernier shown in Fig. 4.7, the first thing to notice is that 25 vernier divisions are equal in length to 24 main scale divisions, each main scale division being 0·50 mm. Thus the instrument can be read to an accuracy of 0·02 mm.
(1) Read off the number of millimetres up to the zero line on the vernier. This is 31·5 mm.
(2) Find the mark on the vernier scale which coincides with a mark on the main scale (this is in Fig. 4.7). Thus, the vernier scale has moved 9×0·02 = 0·18 mm beyond the 31·5 mm mark.

The complete reading is 31·5+0·18 = 31·68 mm.

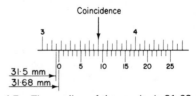

Fig. 4.7   The reading of the vernier is 31·68 mm

Example. The divisions on the main scale of a vernier caliper are 0·5 mm apart. The vernier has 100 divisions equal in length to 98 main scale divisions. To what accuracy can the instrument be read?
    98 main scale divisions = 98×0·5 = 49 mm
    Each division on the vernier scale = $\frac{49}{100}$ = 0·49 mm
    Difference between a main scale division and a vernier division = 0·5−0·49 = 0·01 mm
The instrument can be read to an accuracy of <u>0·01 mm</u> (or $\frac{1}{100}$ of a millimetre).

The vernier height gauge is used for accurate measuring. It is also used when accurate marking out is required. A typical instrument is shown in Fig. 4.8 (a).

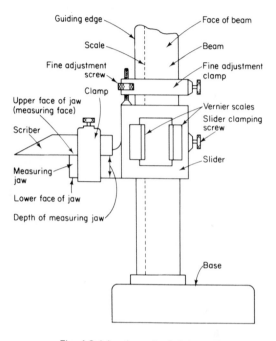

Fig. 4.8 (a)   A vernier height gauge

With new instruments it is possible to measure heights accurately from the surface table. However, after a time the base becomes worn and damaged and measurements taken in this way are unreliable. It is better to rest the work on parallels and measure the difference in height between the parallels and the work as shown in Fig. 4.8 (b).

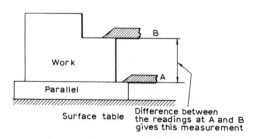

Fig. 4.8 (b)   Use of a vernier height gauge

The vernier depth gauge is used for accurately measuring the depths of holes and recesses as shown in Fig. 4.9.

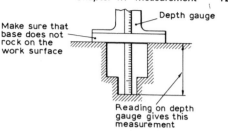

Fig. 4.9   Using a vernier depth gauge

Limits. It is impossible to make anything to an *exact size.* One way of getting over this difficulty is shown in Fig. 4.10 where two dimensions have been placed on

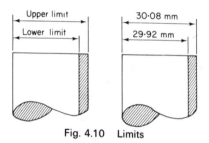

Fig. 4.10   Limits

the drawing. The part is satisfactory if it is not larger than 30·08 mm nor smaller than 29·92 mm. The two dimensions *limit* the size of the part and hence they are called *limits.* The larger dimension is called the upper limit and the smaller dimension the lower limit. Fig. 4.11 shows several ways in which limits are placed on engineering drawings.

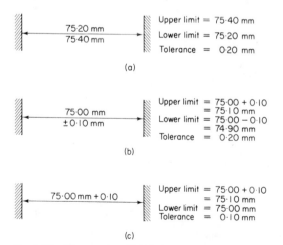

Fig. 4.11   The ways in which limits are placed on drawings

Nominal size. The hole shown in Fig. 4.12 has a nominal diameter of 50·00 mm. According to the drawing the diameter may be between 50·10 mm and 49·90 mm but it will be said to be a 50·00 mm diameter hole.

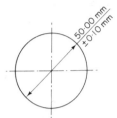

Fig. 4.12   The nominal size is 50·00 mm

Tolerances. The tolerance is the amount of error that can be allowed for imperfect workmanship. The tolerance is obtained by subtracting the lower limit from the upper limit. Thus,

in Fig. 4.11 (a)   Tolerance = 75·40−75·20 = 0·20 mm

in Fig. 4.11 (b)   Tolerance = 75·10−74·90 = 0·20 mm

in Fig. 4.11 (c)   Tolerance = 75·10−75·00 = 0·10 mm

Tolerances may be either unilateral or bilateral. If the tolerance is *unilateral* the limits are on one side of the nominal size, for example $12·00^{+0·02}_{+0·01}$ mm. If the tolerance is *bilateral* the limits are stated on each side of the *nominal size*, for example 12·00±0·10 mm or $12·00^{+0·10}_{-0·08}$ mm.

Types of fit. When a hole and a shaft are to fit together two kinds of fit are possible:

1. Clearance fit which occurs when the shaft is smaller than the hole (Fig. 4.13).

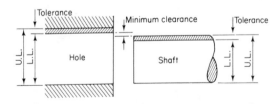

Fig. 4.13   A clearance fit between a hole and shaft. The limits on the hole and the shaft ensure that the shaft is always smaller than the hole

2. Interference fit which occurs when the shaft is larger than the hole (Fig. 4.14).

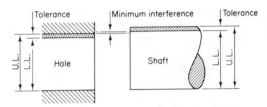

Fig. 4.14   An interference fit between a hole and shaft. The limits on the hole and shaft make sure that the shaft is always larger than the hole

3. Transition fit. The amount of clearance or interference depends upon the limits placed on the hole and shaft. It is possible to arrange the limits so that either a clearance or an interference fit is obtained, depending on the final sizes to which the hole and shaft are made.

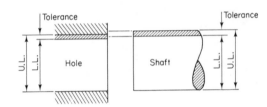

Fig. 4.15   A transition fit between a hole and shaft. When the hole is made to the upper limit and the shaft is made to the lower limit the result is a clearance fit. When the shaft is made to the upper limit an interference fit results

Such an arrangement is called a transition fit (Fig. 4.15). It must be clearly understood that when the shaft and hole are mated together either a clearance fit or an interference fit is obtained. Transition fits are sometimes called *push fits.*

Example. The limits shown on a drawing for a mating hole and shaft are:

For the hole: $10·000^{+0·018}_{+0·000}$ mm

For the shaft: $10·000^{-0·006}_{-0·017}$ mm

Find what type of fit will exist between the hole and shaft.

The smallest hole is 10·000+0·000 = 10·000 mm

The largest shaft is 10·000−0·006 = 9·994 mm

Hence the shaft is always *smaller* than the hole and a *clearance fit* will be obtained.

Example. The limits shown on a drawing for a mating hole and shaft were:

For the hole: $30 \cdot 000^{+0 \cdot 025}_{+0 \cdot 000}$ mm

For the shaft: $30 \cdot 000^{+0 \cdot 018}_{+0 \cdot 002}$ mm

Find the type of fit that exists between the hole and shaft.

The largest shaft is $30 \cdot 000 + 0 \cdot 018 = 30 \cdot 018$ mm

The smallest hole is $30 \cdot 000 + 0 \cdot 000 = 30 \cdot 000$ mm

This combination gives an *interference fit.*

The smallest shaft is $30 \cdot 000 + 0 \cdot 002 = 30 \cdot 002$ mm

The largest hole is $\quad 30 \cdot 000 + 0 \cdot 025 = 30 \cdot 025$ mm

This combination gives a *clearance fit.*

Hence, since we can have either interference or clearance, we have a *transition fit.*

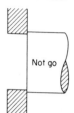

Hole is larger than the lower limit. The gauge goes into the hole. The hole is satisfactory at the lower limit

Hole is smaller than the upper limit. The gauge will not go into the hole. The hole is satisfactory at the upper limit

Example. The limits shown on drawing for a mating hole and shaft were:

For the hole $40 \cdot 000^{+0 \cdot 025}_{+0 \cdot 000}$ mm

For the shaft: $40 \cdot 000^{+0 \cdot 059}_{+0 \cdot 043}$ mm

Find the type of fit that exists between the hole and shaft.

The largest hole is $40 \cdot 000 + 0 \cdot 025 = 40 \cdot 025$ mm

The smallest shaft is $40 \cdot 000 + 0 \cdot 043 = 40 \cdot 043$ mm

The shaft is always *larger* than the hole and hence we have an <u>interference fit.</u>

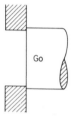

Hole is smaller than the lower limit. The gauge will not go into the hole. Hence the hole is unsatisfactory at the lower limit

Hole is larger than the upper limit. The gauge goes into the hole. Hence the hole is unsatisfactory at the upper limit

Limit gauges. When checking work made to limits, limit gauges are usually used. We check the accuracy of the work by *comparing* the gauge with the work. Thus, there is a distinction between gauging and measuring. When we measure a part we use an instrument (for example a micrometer), which tells us its size in millimetres or inches. When we use a limit gauge all that we know is that the part is either satisfactory or unsatisfactory. We do not know the size to which the part has been made.

The principle of limit gauging is shown in Figs 4.16 and 4.17.

Fig. 4.18 shows a taper-plug gauge for checking a taper hole whilst Fig. 4.19 shows a taper-ring gauge for checking tapered shafts and spindles.

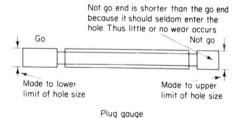

Not go end is shorter than the go end because it should seldom enter the hole. Thus little or no wear occurs

Go   Not go

Made to lower limit of hole size   Made to upper limit of hole size

Plug gauge

Fig. 4.16   Principle of hole gauging

Exercise 4.1

1. State the accuracy to which measurements may be made when using: (1) a micrometer; (b) calipers; (c) a rule.
2. State the readings for the micrometer settings shown in Fig. 4.20.
3. State the readings on the verniers shown in Fig. 4.21.
4. The divisions on the main scale of a vernier instrument are 1 mm apart. The vernier has 50 divisions equal in length to 49 main scale divisions. To what accuracy can the instrument be read?
5. The screw thread for a micrometer has a pitch of 0·5 mm. How many turns are needed to move the thimble 8 mm?
6. Sketch or perform the micrometer settings for the following readings: (a) 8·68 mm; (b) 12·32 mm; (c) 19·79 mm.
7. State instruments suitable for measuring:
   (a) A bar 150 mm diameter to within 0·06 mm;
   (b) The diameter of a bar which is 60 mm to within 0·10 mm;

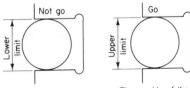

The not go side of the gap gauge is made so that the gap is equal to the lower limit

The go side of the gap gauge is made so that the gap is equal to the upper limit

The shaft is made within limits

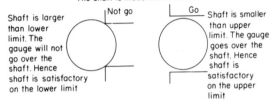

Shaft is larger than lower limit. The gauge will not go over the shaft. Hence shaft is satisfactory on the lower limit

Shaft is smaller than upper limit. The gauge goes over the shaft. Hence shaft is satisfactory on the upper limit

The shaft is outside limits

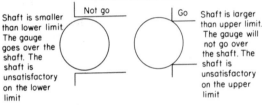

Shaft is smaller than lower limit. The gauge goes over the shaft. The shaft is unsatisfactory on the lower limit

Shaft is larger than upper limit. The gauge will not go over the shaft. The shaft is unsatisfactory on the upper limit

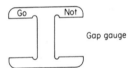

Gap gauge

Fig. 4.17   Principle of shaft gauging

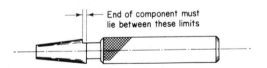

End of component must lie between these limits

Fig. 4.18   Taper-plug gauge

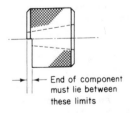

End of component must lie between these limits

Fig. 4.19   Taper-ring gauge

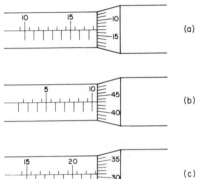

(a)

(b)

(c)

(d)

Fig. 4.20

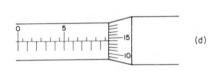

(a)

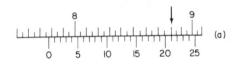

(b)

Coincidence

(c)

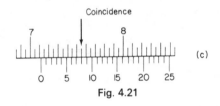

Fig. 4.21

(c)   The length of a bar 220 mm long to within 0·5 mm;

(d)   A bar 75 mm diameter to within 0·8 mm.

8.   A vernier height gauge is being used to mark out the centre-line of a rectangular plate which is mounted on parallels. The following readings are obtained:

   Reading over parallel strip = 34·90 mm
   Reading over top of plate = 204·98 mm

   Find the vernier setting for the centre height of the plate.

9.   The dimension of a shaft is given as 40·00 ± 0·15 mm. Find the upper limit, the lower limit and the tolerance.

10. A hole is stated as being 50.00 ± 0.10 mm diameter. What is the nominal size of the hole?

11. The diameter of a shaft is $15.00^{+0.03}_{+0.01}$ mm. Is the tolerance unilateral or bilateral?

12. State the conditions needed for: (a) a clearance fit; (b) an interference fit for a hole and shaft assembly.

13. By means of a clear sketch show what is meant by a transition fit.

14. The following limits refer to mating holes and shafts. For each case state the type of fit (clearance, interference or transition) and calculate the greatest and least clearance or interference (all dimensions are in millimetres):

    (a) $50.00^{+0.03}_{-0.02}$ for the hole

          $50.00^{+0.10}_{+0.08}$ for the shaft

    (b) $58.00^{+0.03}_{-0.01}$ for the hole

          $58.00^{+0.03}_{-0.02}$ for the shaft

    (c) $12.00^{+0.01}_{+0.00}$ for the hole

          $12.00^{-0.03}_{-0.05}$ for the shaft

    (d) $100.00^{+0.03}_{+0.02}$ for the hole

          $100.00^{-0.05}_{-0.10}$ for the shaft

    (e) $35.00^{+0.03}_{-0.12}$ for the hole

          $35.00^{-0.02}_{-0.04}$ for the shaft

    (f) $25.00^{+0.02}_{+0.00}$ for the hole

          $25.00^{+0.01}_{-0.01}$ for the shaft

    (g) $50.00^{+0.03}_{+0.00}$ for the hole

          $50.00^{-0.05}_{-0.10}$ for the shaft

    (h) $18.00^{+0.02}_{+0.00}$ for the hole

          $18.00^{+0.01}_{+0.00}$ for the shaft

4.2 Marking out

Principle of marking out. The purposes of marking out are:

(a) To define the shape of the outline of the article.
(b) To indicate exactly the positions of holes and similar features.

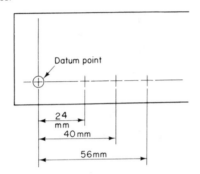

Use of a datum point

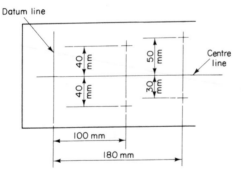

Use of a datum line and a centre line

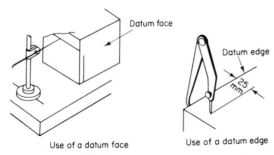

Use of a datum face   Use of a datum edge

Fig. 4.22   All measurements must start from a datum

(c) To make marks which will allow a machinist to set up the work correctly on his machine and to serve as a guide that the correct size has been attained.
(d) When metal has to be removed from several faces marking out ensures that the correct amount is removed from each face.
(e) To keep the wastage of material to a minimum.

All measurements must start from a datum (see Fig. 4.22), which may be a point, an edge, a line or a face. It is more accurate to mark each dimension from the datum than marking off from one position to the next.

To mark out the positions of holes we always need two datums (Fig. 4.23), which are usually at right-angles to each other. In many cases we need three datum faces to define the shape of an article (Fig. 4.24).

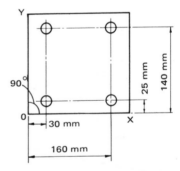

Fig. 4.23   To position the four holes shown two datum edges OX and OY are needed. These edges are used in marking out the holes

Most marking off is performed on a surface plate or surface table (Fig. 4.25), which provides a horizontal surface. A datum face is prepared on the work by machining or by using hand tools and this face is placed on the table. Lines parallel to the surface table are marked by using a scribing block (Fig. 4.26), or a vernier height gauge (Fig. 4.8(a)). By using a datum face at right-angles to the first and placing this on the surface table, lines at right-angles to the first lines may be scribed.

Marking out using a scribing block is unlikely to be closer to size than 0·15 mm. When greater accuracy is needed a vernier height gauge is used, when accuracy up to about 0·02 mm may be achieved.

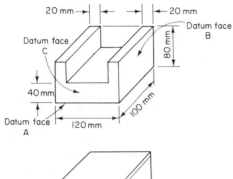

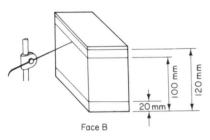

Face A

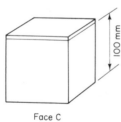

Face B

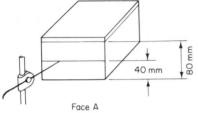

Face C

Fig. 4.24 To position the features of the block the three datum faces A, B and C are needed

Tools and equipment used in marking out (Fig. 4.25). *Dividers* are used for marking out circles between 75 mm and 250 mm. *Jennies* (or odd-legs) are used to scribe lines parallel to an edge, to find centre-lines and to find the centres of round bars. *Trammels* are used for marking out large circles. Marking out using these tools will not be better than about 0·15 mm off the true size.

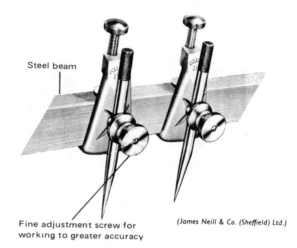

Steel beam

Fine adjustment screw for working to greater accuracy

(James Neill & Co. (Sheffield) Ltd.)

(James Neill & Co. (Sheffield) Ltd.)

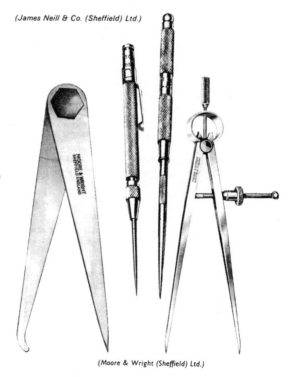

(Moore & Wright (Sheffield) Ltd.)

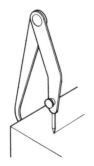

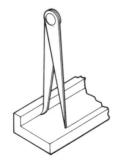

Fig. 4.25 Tools and equipment used in marking out

(*Windley Bros. Ltd.*)

Fig 4.25 (*continued*)

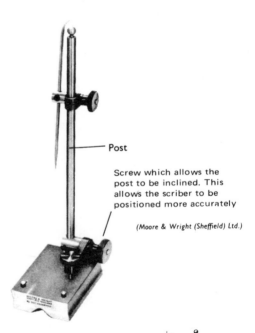

Post

Screw which allows the post to be inclined. This allows the scriber to be positioned more accurately

(*Moore & Wright (Sheffield) Ltd.*)

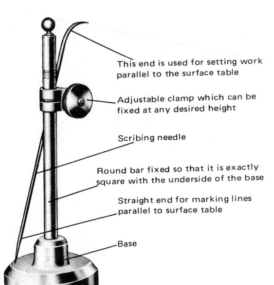

This end is used for setting work parallel to the surface table

Adjustable clamp which can be fixed at any desired height

Scribing needle

Round bar fixed so that it is exactly square with the underside of the base

Straight end for marking lines parallel to surface table

Base

(*Moore & Wright (Sheffield) Ltd.*)

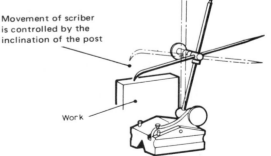

Movement of scriber is controlled by the inclination of the post

Work

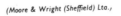

Fig. 4.26   Scribing blocks

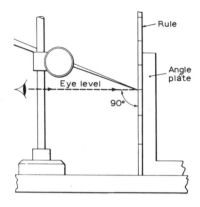

Fig. 4.26 (*continued*)

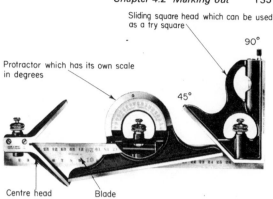

Measurement of angles. For measuring and checking angles of 90° the engineers' try-square (Fig. 4.27) is

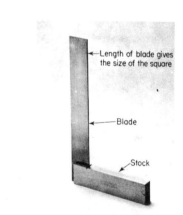

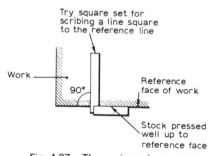

Fig. 4.27   The engineers' try-square

used. It may also be used for marking out lines at right-angles to a datum face or edge.

For checking angles some form of protractor is needed. The combination set (Fig. 4.28) is often used for this purpose.

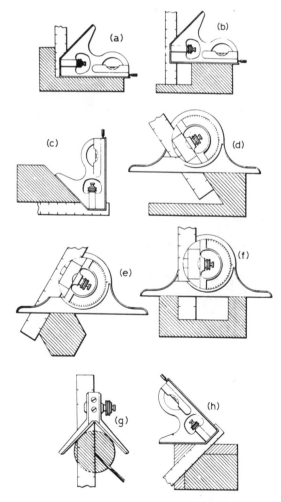

Fig. 4.28   The combination set and its uses

For marking off angles the bevel (Fig. 4.29) is commonly used but it must be set to the correct angle before use.

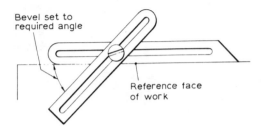

Fig. 4.29   The bevel

Marking off round work. Vee blocks are used to support round work (Fig. 4.30). The vee must be large enough to locate the work properly (Fig. 4.31).

(James Neill & Co. (Sheffield) Ltd.)

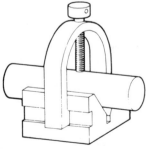

Fig. 4.30   Vee blocks

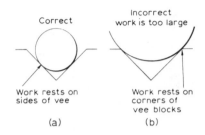

Fig. 4.31   The vee must be large enough to locate the work properly

Plumb-lines and spirit levels. The plumb-line provides a vertical datum line when used as shown in Fig. 4.32. The spirit level (or 'bubble' level) is used to make sure that surfaces are horizontal. Strictly speaking a spirit level checks the angle of tilt of a surface as shown in Fig. 4.33. When the surface is horizontal the bubble positions itself at the centre of the vial. When the surface is tilted the bubble moves away from the centre of the vial. The distance of the bubble away from the centre position is a measure of the angle of tilt of the surface.

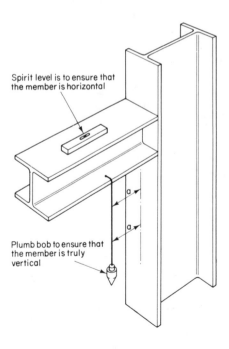

Fig. 4.32   Use of spirit level and plumb bob to make sure that structural members are horizontal and vertical

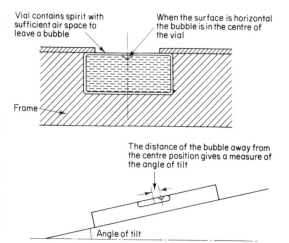

Vial contains spirit with sufficient air space to leave a bubble

When the surface is horizontal the bubble is in the centre of the vial

Frame

The distance of the bubble away from the centre position gives a measure of the angle of tilt

Angle of tilt

Fig. 4.33 The spirit level

Templates. Structures such as roof trusses, bridge girders and cranes are made up of a great many different parts. These parts are made separately in a fabrication shop and transported to the site for assembly and erection. In order to make sure that everything fits properly templates are made for each part. These templates are made from plywood, whitewood, hardboard, millboard or special template-making paper. The templates have holes drilled in them to represent holes in the finished parts and they also have provisions made so that the outlines of the part can be marked. Fig. 4.34 shows a template intended for a plate detail.

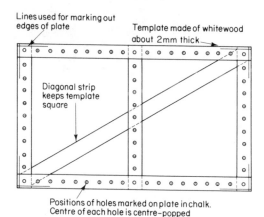

Lines used for marking out edges of plate

Template made of whitewood about 2mm thick

Diagonal strip keeps template square

Positions of holes marked on plate in chalk. Centre of each hole is centre-popped

Fig. 4.34 A wooden template for a plate detail

Some detailed parts of a structure are so simple that they can be marked off directly from drawings at the bench in the fabrication shop. Nevertheless templates for these simple details are used where a number of

identical parts have to be made since marking out is then quicker and more accurate than measuring each detail separately.

Very often holes in the flanges of joists and other structural sections are not dimensioned on the working drawings. Standard dimensions, decided by previous arrangement with the drawing office, are used by the template maker in order to position the holes (Fig. 4.35).

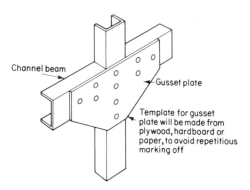

Channel beam

Gusset plate

Template for gusset plate will be made from plywood, hardboard or paper, to avoid repetitive marking off

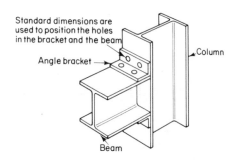

Standard dimensions are used to position the holes in the bracket and the beam

Angle bracket

Column

Beam

Fig. 4.35 Typical structural connections

Exercise 4.2
1. Describe briefly why parts are marked out.
2. What are (a) datum points (b) datum lines (c) datum faces? Give an example where each might be used.
3. How are lines parallel to a datum face marked off?
4. What is the accuracy of the commonly-used marking-off tools?
5. Show how jennies are used for marking out?
6. How may angles be checked?
7. When would an angle plate be used?
8. What is the purpose of vee blocks? When are clamps used with vee blocks?
9. What is a plumb-line used for?
10. How does a spirit level give a measure of the angle of tilt of a surface?
11. State three reasons why templates are used.

4.3   Work and tool holding

The six ways in which movement can take place.
The block (Fig. 4.36) can move in any of the six ways shown. That is, it can slide in three ways and rotate in three ways.

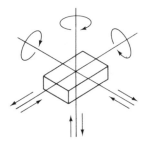

Fig. 4.36   The six ways in which movement can take place

Restricting movement. In order to keep a workpiece in a given position it is necessary to prevent it moving in any of the ways shown in Fig. 4.36. Restriction of movement can occur in one of two ways:
(1) by positive location (Fig. 4.37).
(2) by frictional resistance (Fig. 4.38).

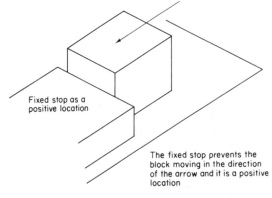

Fixed stop as a positive location

The fixed stop prevents the block moving in the direction of the arrow and it is a positive location

Fig. 4.37   Positive location

In practice workpieces are held by a combination of positive location and friction. The main cutting force should always be restrained by a positive location. Now let us see how these principles are applied for some of the more common work- and tool-holding devices.

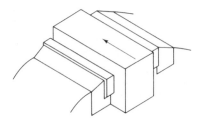

The frictional resistance between the work and the faces of the vice tends to prevent the work from moving in the direction of the arrow

Fig. 4.38   Frictional resistance

1. *The bench vice*. The bench vice (Fig. 4.39) has a fixed jaw and a moving jaw. The fixed jaw is used as the positive location and hence it should be used to restrain the main cutting force.

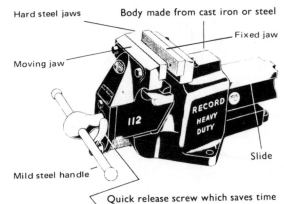

Hard steel jaws   Body made from cast iron or steel

Fixed jaw

Moving jaw

112   RECORD HEAVY DUTY

Slide

Mild steel handle

(C. & J. Hampton Ltd.)

Quick release screw which saves time when the jaws have to be moved a considerable distance.

Fig. 4.39   Bench vice

Fig. 4.40 shows how movement of the work is prevented when using a bench vice. Note that the greater the force holding the work, the greater the frictional resistance. Therefore make sure that the vice is well tightened before starting to cut. When this is done the hard steel jaws of the vice may damage light and polished work. Vice clamps (Fig. 4.41) are used to prevent this.
2. *Vee blocks*. These are used for supporting round

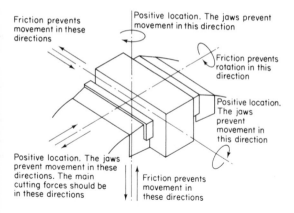

Friction prevents movement in these directions

Positive location. The jaws prevent movement in this direction

Friction prevents rotation in this direction

Positive location. The jaws prevent movement in this direction

Positive location. The jaws prevent movement in these directions. The main cutting forces should be in these directions

Friction prevents movement in these directions

Fig. 4.40   How movement of the work is prevented when using a bench vice

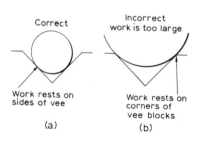

Correct

Incorrect work is too large

Work rests on sides of vee

Work rests on corners of vee blocks

(a)

(b)

Fig. 4.43

When a clamp is used (Fig. 4.44), movement is prevented in the four directions which were not restrained previously.

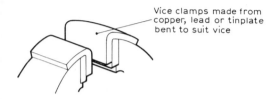

Vice clamps made from copper, lead or tinplate bent to suit vice

Fig. 4.41   Vice clamps are used to prevent damage to light and polished work

work. They are supplied in pairs so that each is a perfect match. The centre vee is 90°. Fig. 4.42 shows that work placed in a vee block is free to move in four directions (one rotational and three in a straight

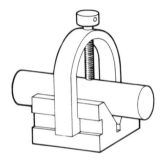

Fig. 4.44

3. *Clamping.* Work often has to be clamped to prevent it moving under the action of cutting forces. Fig. 4.45 shows the principles of location and restraint when clamping. Note that the bolt provides the force needed to give the frictional resistance.

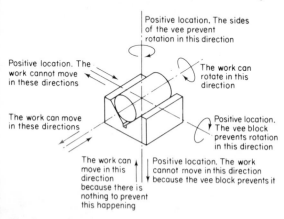

Positive location. The sides of the vee prevent rotation in this direction

Positive location. The work cannot move in these directions

The work can rotate in this direction

The work can move in these directions

Positive location. The vee block prevents rotation in this direction

The work can move in this direction because there is nothing to prevent this happening

Positive location. The work cannot move in this direction because the vee block prevents it

Fig. 4.42   How the more important movements are prevented by using a vee block

line). Note also (Fig. 4.43) that the work must rest on the sides of the vee for the vee block to support the work properly.

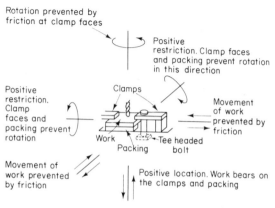

Rotation prevented by friction at clamp faces

Positive restriction. Clamp faces and packing prevent rotation in this direction

Positive restriction. Clamp faces and packing prevent rotation

Clamps

Movement of work prevented by friction

Work

Packing

Tee headed bolt

Movement of work prevented by friction

Positive location. Work bears on the clamps and packing

Fig. 4.45   Work clamped to prevent motion when drilling it

Exercise 4.3
For each of the diagrams (Fig. 4.46) state how movement
is prevented in the direction shown.

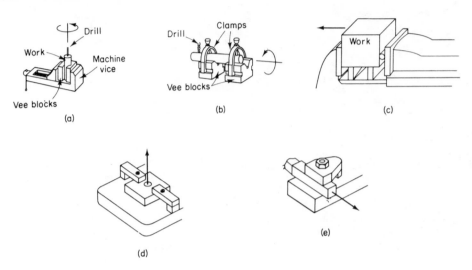

(a)   (b)   (c)

(d)

(e)

Fig. 4.46

4.4   Cutting tools

Tool materials. For a tool to cut successfully the material from which the tool is made must be harder than the material which is being cut. The materials which are available for cutting tools have been discussed in Chapter 2.8.

The tool or point angle. Fig. 4.47 shows two kinds of tool points. The one at (a) is too weak for it to be able to cut metal successfully—the edge would soon crumble away. On the other hand, the tool point at (b) is strong and robust. The greater the tool angle the stronger the tool.

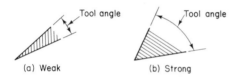

Fig. 4.47   The point or tool angle

The strong point at Fig. 4.47 (b) needs a great deal of effort to force it through the metal, whilst that at (a) requires considerably less effort. Hence, we always have to compromise between strength and the effort needed for cutting.

The clearance angle. Fig. 4.48 (a) shows a cold chisel taking a continuous cut. The energy needed to cut the metal is supplied by the hammer blow. Fig. 4.48 (b) shows an enlarged drawing of the chisel point. The heel of the

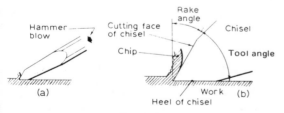

Fig. 4.48

chisel is flat on the work, this ensuring that the depth of cut is maintained. As the chisel moves along the work the heel rubs. However since the chisel moves slowly this rubbing action is small and not much effort is needed to keep the chisel moving. Many tools, however, move at a very fast speed and this rubbing action becomes very important. Clearance between the work and the tool

must be provided (Fig. 4.49) because friction and heat will cause the tool to wear out quickly. The clearance angle on a tool is kept as small as possible (usually between 5° and 7°). If the clearance angle is made too large the tool point will be weakened and the tool may 'dig in' thus producing a poor finish on the work.

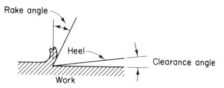

Fig. 4.49   Clearance angle to prevent the heel of the tool rubbing on the work

Rake angle. It will be seen from Fig. 4.48 that the metal being cut moves up the cutting face of the chisel. The angle that the cutting face of a tool makes with a line perpendicular to the work surface is called the *rake angle*. The rake angle is very important indeed. When ductile metals are to be cut a large rake angle makes cutting easier and less effort will be expended in cutting. However, if the rake angle is made too large the tool point will be considerably weakened and a compromise must again be used. Fig. 4.50 shows all the tool angles previously discussed.

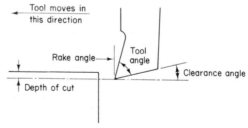

Fig. 4.50   The angles on a cutting tool

Chip formation. There are three fundamental types of chip which are produced when metal is cut.
1. *Discontinuous chip* which is produced when brittle metals like cast iron are being cut. These chips appear as small pieces of metal or granules as shown in Fig. 4.51. There is little or no flow or metal over the face of the tool and hence when cutting brittle metals little or no rake angle is needed.

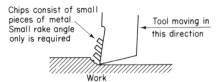

Chips consist of small pieces of metal. Small rake angle only is required

Tool moving in this direction

Work

Fig. 4.51   Discontinuous chips are produced when brittle metals are cut

2. *Continuous chip* which is obtained when a ductile metal like mild steel is being cut. The chip is ribbon-like and flows over the tool face as shown in Fig. 4.52. Thus, when cutting ductile metals the rake angle should be as large as possible so that the chip

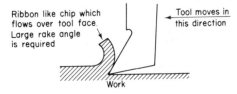

Ribbon like chip which flows over tool face. Large rake angle is required

Tool moves in this direction

Work

Fig. 4.52   Continuous chips are produced when ductile metals are cut

can flow easily. The metal is sheared off the work just ahead of the cutting edge of the tool, leaving a rough finish on the work. This rough finish is then smoothed off by the sharp cutting edge of the tool. A blunt tool will cut the metal and form a chip but the work will be left rough (i.e. a poor finish will be obtained). Very soft materials such as copper tear too deeply for even a sharp tool to smooth. It is very difficult to produce a good finish on soft, ductile metals.

3. *Continuous chip with chip welding.* As previously mentioned, the cutting edge of a tool smoothes off the torn surface of the work. Thin pieces of metal are produced which become trapped between the chip and the face of the tool. The combination of pressure and heat causes these small pieces of metal to become welded to the tip of the tool thus forming a built-up edge (Fig. 4.53). A built-up edge can be avoided by:

(a) Reducing friction by giving the face of the tool a very fine finish.

(b) Using a lubricant to prevent the small pieces of metal adhering to the tool. The fluid also helps by flushing out these small pieces of metal.

(c) Reducing the pressure. Increasing the rake angle and reducing the chip thickness by using a reduced feed rate causes the pressure to be reduced.

(d) Using a free-cutting metal for the work.

Effect of cutting speed. When metal is cut at a high speed a great deal of heat is generated at the cutting edge of the tool. If this heat cannot be got rid of quickly enough the tool softens because of the high temperature and soon becomes blunt. For this reason it is unwise to exceed the

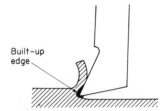

Built-up edge

The built-up edge forms on the face of the tool. As more metal welds on, the built-up edge protrudes beyond the cutting edge

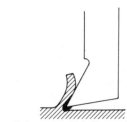

The chip squeezes the built-up edge onto the clearance face of the tool

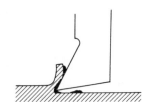

Pieces of the built-up edge become detached and adhere to the work giving it a rough finish. Some becomes attached to the chip

Fig. 4.53   The built-up edge and its effects

cutting speed recommended for the cutting tools and the materials being cut.

Effect of feed. Generally speaking a fine feed will give a good surface finish. It is possible to have too fine a feed in which case the tool may not cut cleanly. It usually pays to take a deep cut at a low feed rate. Doing this gives a good finish and removes metal more quickly than a shallow cut at a high feed rate.

Power used in cutting. When cutting a ductile metal like mild steel it is the rake angle on the tool which determines the amount of power which will be used. Fig. 4.54 shows the effect of rake angle on power. It will be noticed that a large rake angle reduces the power required. However, by increasing the rake angle too much the tool point is weakened and a compromise is used (Fig. 4.55).

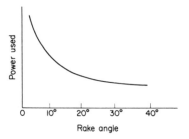

Fig. 4.54   Rake angle

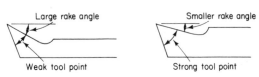

Fig. 4.55   The effect of rake angle on the strength of the tool point

Grinding of tools. As a tool cuts, it loses its keen cutting edge and becomes dulled. The tool then needs resharpening. Grinding wheels are generally used for this purpose. Artificial abrasives are used in their manufacture and the best abrasive for metal-working tools is:

Aluminium oxide. The grains of the abrasive are held in a bonding medium. When the wheel is used the abrasive grains become blunted. For the wheel to cut properly these blunted grains must be shed so that fresh, sharp

grains can take their place. The bonding material must be such that the new, sharp grains are securely held, whilst the blunted grains are removed by the pressure of the work on the wheel. The wheels are mounted on a tool-grinding machine which is usually provided with a tool rest.

Exercise 4.4

1. In Fig. 4.56 state the tool angle, the clearance angle and the rake angle.

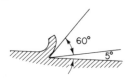

Fig. 4.56

2. What is the purpose of the clearance angle on a tool?
3. Why is it important that the correct rake angle is used on a cutting tool?
4. Why is the clearance angle on a tool kept to about 5°?
5. Name the three types of chip which may be produced when cutting metal. Name a metal where each is likely to occur.
6. Why is it important to use the correct cutting speed for a tool?
7. How is a good finish likely to be obtained?
8. How does the rake angle on a tool affect the power used?

4.5 Small tools

Hammers (Fig. 4.57). The masses of hammers vary

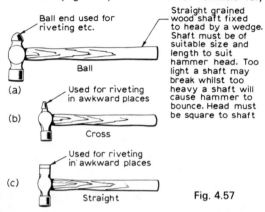

Ball end used for riveting etc.

Straight grained wood shaft fixed to head by a wedge. Shaft must be of suitable size and length to suit hammer head. Too light a shaft may break whilst too heavy a shaft will cause hammer to bounce. Head must be square to shaft

(a) Ball

(b) Used for riveting in awkward places

Cross

(c) Used for riveting in awkward places

Straight

Fig. 4.57

from about 100 grams to 1500 grams. Hammers between 100 and 250 grams are used for centre punching etc. A 500 gram hammer is used for driving pins, etc., whilst a 1000 gram (1 kg) hammer is used for chiselling. A hammer of 1500 grams (1½ kg) is used for heavy work.

Soft hammers are used where the work will be damaged if an ordinary hammer is used.

The faces of soft hammers must be kept clean, embedded metal chips cause damage to the work. Do not use a soft hammer on rough work as it will become damaged.

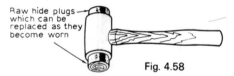

Raw hide plugs which can be replaced as they become worn

Fig. 4.58

Files. The teeth of files may be either single cut or double cut.

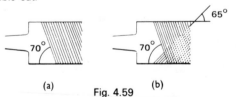

70° (a)

65° 70° (b)

Fig. 4.59

(a) Single cut file used for soft metals.
(b) Double cut file used for most metal work.

Grades of files. The grade depends upon the length of the file. A short file will have more teeth per centimetre than a long file. The table below gives details and uses of the various grades of file.

Grade	Number of teeth per centimetre	Use
Rough	5–8	Soft metals and plastics.
Bastard	6–16	Iron castings and general roughing-out.
Second cut	7–17	Roughing-out hard metals and finishing soft metals.
Smooth	12–25	Finishing cuts and draw filing.

Types of files. Fig. 4.60 shows the basic shape of a file. The types of files used for various jobs are shown in Fig. 4.61.

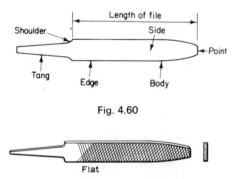

Length of file

Shoulder

Side

Point

Tang

Edge

Body

Fig. 4.60

Flat

A file of rectangular cross-section parallal for two-thirds of the body length and then tapering towards the point in width. The sides are double cut and the edges single cut. Used for all kinds of flat surfaces.

Hand

A file of rectangular cross-section parallel in width over the complete body length. It is parallel in thickness for about two-thirds of the body length and then tapers towards the point. The sides are double cut whilst one edge is single cut, the other edge being left uncut. Used for flat surfaces adjacent to shoulders and for parallel slots.

Square

A file of square cross-section parallel for two-thirds of the body length and then tapering towards the point. All sides are double cut. Used for finishing square and rectangular holes and narrow slots.

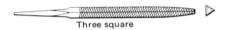

Three square

A file of equilateral triangular cross-section parallel for two-thirds of the body length and then tapering towards the point. All the sides are double cut. Used for sharp corners, internal angles and for making marks to define positions.

Fig. 4.61   Types of files

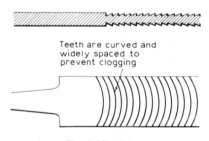

Teeth are curved and widely spaced to prevent clogging

Fig. 4.62   Rasp

Rasps are used for filing soft metals such as aluminium.

Use of files

Cleaning files. When filing, the teeth of the file become clogged or 'pinned' with small pieces of metal. A wire brush or a file card is used to remove the pinning. To prevent pinning, chalk is sometimes rubbed on the file. Turpentine is used for the same purpose when filing aluminium.

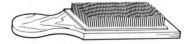

Fig. 4.63   A file card

File handles.
(1) Use the correct size of handle. A large file needs a large handle; a small file needs a much smaller handle.
(2) Never use a split handle as this is dangerous.
(3) Using one handle for several files is bad practice. Once a handle has been fitted to a file it should remain until the file is worn out.

Stages of use of files. New files should be used first on copper, aluminium, brass, bronze and similar metals, and then in turn on cast iron, wrought iron and steel. A new file must never be used on a metal of unknown hardness, especially the surface of a casting. Only the oldest files should be used for such jobs.

Care of files. Files should be stored each in its own wooden rack, and never heaped together. Never strike a file with a hammer or throw it on to metal articles; otherwise broken teeth may result.

Cold chisels

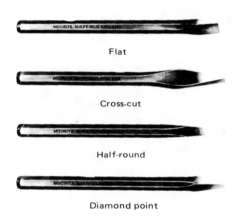

Flat

Cross-cut

Half-round

Diamond point

Fig. 4.64   Cold chisels

The *flat chisel* is used for cutting sheet metal, cutting grooves and chiselling flat surfaces.

The *cross-cut chisel* is used for cutting narrow grooves and keyways.

The *half-round chisel* is used for forming oil channels in bearings, etc.

The *diamond-point chisel* is used for clearing corners and cutting vee grooves, etc. As the point is weak only light blows should be used.

Cutting angles for a cold chisel. When a chisel is inclined for cutting (Fig. 4.65) the rake angle is easily obtained from the point angle and the angle of inclination of the chisel as follows:

Rake angle = $90° - (\text{angle of inclination} + \frac{1}{2} \times \text{point angle})$

CHISEL POINT ANGLES

Metal being cut	Point angle	Angle of inclination
High-carbon steel	65°	39½°
Cast iron	60°	37°
Mild steel	55°	34½°
Brass	50°	32°
Aluminium	30°	22°

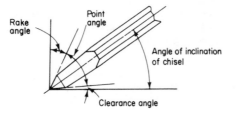

Fig. 4.65   Chisel cutting angles

Example. When cutting brass a chisel with a point angle of 50° is inclined at 32° to the work. What is the rake angle and what is the clearance angle?

Clearance angle = angle of inclination$-\frac{1}{2}\times$point angle
= 32°$-\frac{1}{2}\times$50°
= 32°$-$25° = <u>7°</u>

Rake angle   = 90°$-$(angle of inclination$+\frac{1}{2}\times$point angle)
= 90°$-$(32°$+\frac{1}{2}\times$50°)
= 90°$-$(32°$+$25°)
= 90°$-$57° = <u>33°</u>

Hacksaw teeth

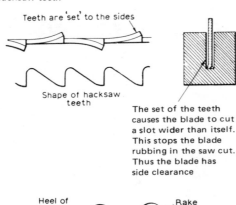

Teeth are 'set' to the sides

Shape of hacksaw teeth

The set of the teeth causes the blade to cut a slot wider than itself. This stops the blade rubbing in the saw cut. Thus the blade has side clearance

Heel of tooth

Rake angle

Work

Clearance angle

Fig. 4.66   Hacksaw teeth

Hacksaws

Types of blades. Two kinds are used—the 'all-hard' and the flexible. The all-hard blade is more easily broken but it allows for more accurate work. The flexible blade is useful when the work is awkwardly placed.

The clearance angle on the tooth prevents the heel from rubbing on the work. The hacksaw shown in Fig. 4.66 has a rake angle but with fine-toothed saws this tends to weaken the blade. Hence fine-toothed blades have no rake angle provided (Fig. 4.67). The space between the teeth prevents the teeth from becoming clogged with chips.

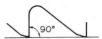

Fig. 4.67   Hacksaw blade with no rake angle

At least three teeth must be in contact with the work-piece or the teeth may be broken off (Fig. 4.68).

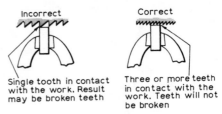

Incorrect

Correct

Single tooth in contact with the work. Result may be broken teeth

Three or more teeth in contact with the work. Teeth will not be broken

Fig. 4.68

Selection of blades. The number of teeth per centimetre varies from 5 to 12 depending on the type of work and type of metal being cut. For soft metals use a coarse blade and for hard metals use a finer blade. The following table will serve to give a guide to blade selection.

Material	Solid metal	Tube and thin sheet
Iron and steel	6–7 teeth/cm	12 teeth/cm
Non ferrous metals	5–6 teeth/cm	8–10 teeth/cm

A speed of 40–50 strokes per minute is fast enough when sawing.

Scrapers. Machined and filed surfaces may look smooth but they actually consist of high spots and low areas. When two such surfaces slide over each other the high spots wear away. The accuracy of fit between the two surfaces is lost. To prevent this occurring the high spots are removed by scraping.

Three basic kinds of scraper are used as shown in Fig. 4.69. Scraping can only be done properly if the cutting edge is kept razor sharp. Hence the blade is constantly touched-up on an oilstone.

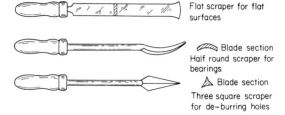

Flat scraper for flat surfaces

Blade section
Half round scraper for bearings

Blade section

Three square scraper for de—burring holes

Fig. 4.69   Types of scrapers

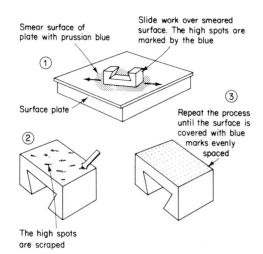

Smear surface of plate with prussian blue

Slide work over smeared surface. The high spots are marked by the blue

①

Surface plate

②

The high spots are scraped

③

Repeat the process until the surface is covered with blue marks evenly spaced

Fig. 4.70   Obtaining a smooth surface by scraping it and comparing it with a reference surface

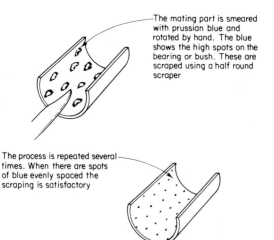

The mating part is smeared with prussian blue and rotated by hand. The blue shows the high spots on the bearing or bush. These are scraped using a half round scraper

The process is repeated several times. When there are spots of blue evenly spaced the scraping is satisfactory

Fig. 4.71   Scraping a bearing. The mating part is used as the reference surface

To find the high spots on the work a reference surface is used. For flat surfaces a surface plate is used. We then *compare* the smoothness of the surface plate with the smoothness of the work by the method shown in Fig. 4.70. When scraping a bearing or a bush the mating part is used as the reference surface (Fig. 4.71).

Safety

(1) Avoid split, broken and loose shafts of hammers. Heads must not be worn or chipped and they must be securely fastened to the shaft.

(2) Files must never be used as levers. They should always have a proper handle fitted to them.

(3) Chisels with burred ends are dangerous. When chipping make sure that the chips do not hit somebody standing nearby.

(4) Always hold the work firmly in a vice or other holding device. If the work shifts during cutting a nasty cut may result.

(5) Cutting tools should be heat-treated by specialists. If the tools are made too hard dangerous fragments may occur when the tools are used.

Exercise 4.5

1. When are soft hammers used in preference to hard-headed hammers? Why must the faces of soft hammers be kept clean?

2. Select suitable grades of file for use on the following metals: (a) aluminium; (b) cast iron; (c) cast steel. What grade should be used for draw filing?

3. State the type of file that would be used for the following purposes: (a) for filing parallel slots; (b) for filing flat surfaces; (c) for filing a rectangular hole.

4. You are given a new file. Write down the order in which you would use the file on the following metals: cast iron, cast steel, brass, mild steel.

5. Name two chisels in common use and give an example of the type of work for which each would be used.

6. When cutting mild steel a chisel with a point angle of 55° is inclined $34\frac{1}{2}°$ to the work. Calculate the rake and clearance angles?

7. Hacksaw blades are available with 5, 7, 9 and 12 teeth per centimetre. State which would be used for cutting: (a) thin-walled tubing; (b) mild steel; (c) aluminium.

8. When is a chisel used without clearance?

9. Why is it essential that a hacksaw tooth should possess clearance.

10. What is meant by the 'set' of hacksaw teeth? What is the purpose of this set?

11. What precautions should be taken when sawing thin-walled tubes?

12. Name three kinds of scrapers in general use and state the purpose for which each would be used.

13. How is a flat surface obtained by scraping?

14. State how a bearing is scraped.

4.6 Drills

Drilling is the operation of making a hole in a material. It must not be confused with boring which is the operation of enlarging a hole which has been previously made. A drill does not produce a very accurate hole. When an accurate hole is required the drilling operation must be followed by reaming or boring.

The twist drill (Fig. 4.72) is the most widely used tool for drilling holes.

Drill angles. For a drill to cut properly it must have a clearance angle and a rake angle. The clearance angle is provided as shown in Fig. 4.72. The rake angle is provided by the helix of the flutes (Fig. 4.73). It is possible to alter the clearance angle when grinding the point but the rake angle is fixed at the time the drill is made. We can, however, choose drills with the correct helix angle for the material being cut. The table opposite gives suitable point and clearance angles and recommended helixes for cutting various materials.

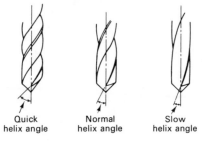

Fig. 4.73 The helix angle

Quick helix angle Normal helix angle Slow helix angle

The drill point. For the drill to cut properly it is essential:
(1) for the lip lengths to be equal;
(2) for the point angle to be the same on each side of the drill centre-line.

The effects of unequal angles and unequal lip lengths are shown in Fig. 4.74. Note that the hole in both cases is larger than the drill.

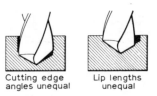

Cutting edge angles unequal   Lip lengths unequal

Fig. 4.74

The web of the drill. Fig. 4.75 shows the web of a drill and how it may be thinned.

This part of the drill is ground away to leave the land, thus decreasing the amount of rubbing on the hole

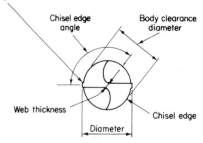

Chisel edge angle

Body clearance diameter

Web thickness

Chisel edge

Diameter

Diameter across lands represents the drill size

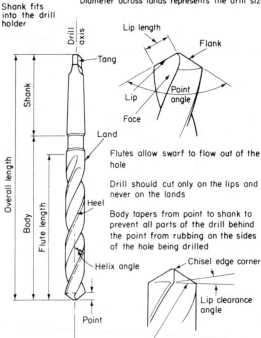

Shank fits into the drill holder

Drill axis

Tang

Shank

Lip length

Flank

Lip

Point angle

Face

Land

Flutes allow swarf to flow out of the hole

Drill should cut only on the lips and never on the lands

Body tapers from point to shank to prevent all parts of the drill behind the point from rubbing on the sides of the hole being drilled

Heel

Helix angle

Overall length

Body

Flute length

Point

Chisel edge corner

Lip clearance angle

Clearance angle to prevent rubbing

Fig. 4.72 The twist drill

Purpose	Point angle	Clearance angle	Type of helix	Type of land	Type of web
General purposes	118°	10–12°	standard	standard	standard
Cast iron	90°	10–12°	standard	standard	standard
Tough steel	130°	12°	slow	standard	thick
Brass, bronze	118°	15°	slow	standard	thin
Copper	100°	15°	quick	narrow	standard
Light alloys	100°	15°	quick	narrow	standard

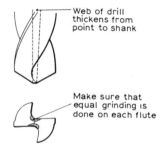

Web of drill thickens from point to shank

Make sure that equal grinding is done on each flute

Fig. 4.75   Shows details of the web

After the drill has been reground several times, the point becomes too thick for efficient cutting. It is then thinned, as shown, on the emery wheel.

Cutting speeds for drills. Drills are different from most other cutting tools in that the cutting speed is zero at the centre of the drill and a maximum at the drill lands. The cutting speeds quoted below are the maximum (i.e. the speed at the drill lands). Most modern drills are made from high-speed steel but sometimes plain carbon steel drills are used.

TABLE OF CUTTING SPEEDS FOR HIGH-SPEED STEEL DRILLS

Metal being drilled	Cutting speed in metres/min
Aluminium	60–76
Brass and bronze	45–60
Cast iron	15–27
Copper	30–45
Mild steel	30–37
High carbon steel	15–20

Carbon steel drills should have cutting speeds of about 60% of those shown in the table.

Feeds for drills. A feed which is too great will cause the drill to break. Recommended feeds for high-speed-steel twist drills are as follows:
under 5 mm diameter 0·040 to 0·150 mm per rev;
between 5 mm and 10 mm diameter 0·150 to 0·250 mm per rev;
over 10 mm diameter 0·250 to 0·500 mm per rev.

Causes of drill failure
(1) A rough hole is usually caused by too rapid a feed or a blunt or incorrectly-ground drill point.
(2) An oversize hole is caused because: (i) the lips of the drill are unequal in length; (ii) the cutting edge angles are unequal; (iii) point thinning is not central.
(3) A split web is the result of: (i) too great a feed; (ii) point being thinned too much; (iii) too small a lip clearance angle.
(4) Damaged corners are caused by: (i) too high a cutting speed; (ii) insufficient coolant.
(5) Causes of drill breakage are: (i) too great a feed; (ii) the drill binding in the hole because the lands are worn away; (iii) the drill becoming choked with swarf; (iv) the point being wrongly ground; (v) the drill slipping in the chuck.

Special tools
1. *The flat drill,* Fig. 4.76, is easy to make. As with the twist drill the two cutting edges and angles must be alike. It is necessary to use a great deal of pressure to force the drill into the work.

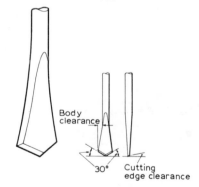

Body clearance

30°   Cutting edge clearance

Fig. 4.76   The flat drill

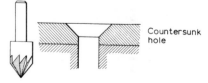

Countersunk hole

Fig. 4.77

2. *Countersinking tools* are used for countersinking holes for screws or rivets. They are made to give either a 90° or a 120° angle of countersink (Fig. 4.77).

3. *Counterboring tools.* Details of counterboring tools and operations are shown in Fig. 4.78.

4. *Multi-flute drills.* The normal twist drill has two flutes. If this type of drill is used to open out a cored hole in a casting (i.e. a hole made during the casting process) it vibrates and a bad hole is produced. A multi-flute drill (i.e. one having more than two flutes) overcomes this objection. This drill is sometimes known as a *core drill*.

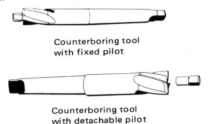

Counterboring tool
with fixed pilot

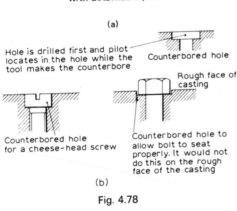

Counterboring tool
with detachable pilot

(a)

Hole is drilled first and pilot
locates in the hole while the
tool makes the counterbore

Counterbored hole

Counterbored hole
for a cheese-head screw

Rough face of
casting

Counterbored hole to
allow bolt to seat
properly. It would not
do this on the rough
face of the casting

(b)

Fig. 4.78

5. *Reamers.* These tools are used when accurately finished holes are needed. The reamer removes only a little metal from the hole and thus brings the drilled hole to size. Although a reamer will produce a hole with a good finish and with the correct size, it will not correct holes which have been incorrectly positioned. The reamer can only follow the drilled hole. Both hand reamers and machine reamers are used.

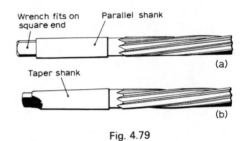

Wrench fits on
square end

Parallel shank

(a)

Taper shank

(b)

Fig. 4.79

(a) Hand reamer.
(b) Machine reamer for use in a drilling machine or a lathe.

Fig. 4.80

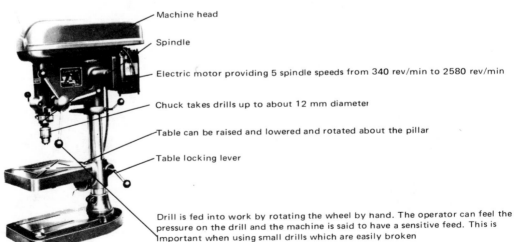

Machine head

Spindle

Electric motor providing 5 spindle speeds from 340 rev/min to 2580 rev/min

Chuck takes drills up to about 12 mm diameter

Table can be raised and lowered and rotated about the pillar

Table locking lever

Drill is fed into work by rotating the wheel by hand. The operator can feel the pressure on the drill and the machine is said to have a sensitive feed. This is important when using small drills which are easily broken

Fig. 4.81   Bench drilling machine   *(B. Elliott (Machinery) Ltd.)*

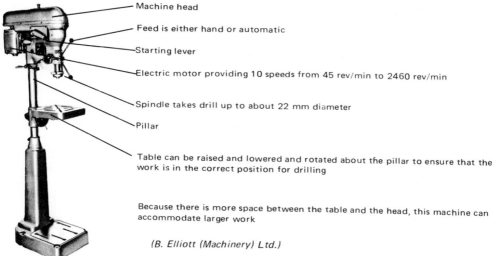

Machine head

Feed is either hand or automatic

Starting lever

Electric motor providing 10 speeds from 45 rev/min to 2460 rev/min

Spindle takes drill up to about 22 mm diameter

Pillar

Table can be raised and lowered and rotated about the pillar to ensure that the work is in the correct position for drilling

Because there is more space between the table and the head, this machine can accommodate larger work

(B. Elliott (Machinery) Ltd.)

Fig. 4.82   Pillar progress machine

Sharpening of twist drills. In grinding twist drills the following errors can occur:
(1) the lips are unequal in length,
(2) the lips are at unequal angles,
(3) the lips are both at unequal angles and unequal in length.

Any of these three errors will cause a bad hole to be drilled. These errors are most likely to occur when the drill is ground by hand. The best way to grind a drill is to use the special drill grinding attachment shown in Fig. 4.80. This attachment ensures that the drill is ground correctly in every way.

Drilling machines. Two types as shown in Figs 4.81 and 4.82 are commonly used. The bench machine is so called because it is small enough to stand on a bench.

Holding the drill. The drill must be firmly held in the drilling machine chuck, Fig. 4.83, so that it is perfectly steady as it rotates. The figure shows a chuck for holding small drills with parallel shanks. It is opened and closed by means of the key. The chuck must be perfectly clean before fitting the drill or the drill will not run true.

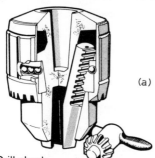

(a)

Fig. 4.83   Drill chuck

When drilling, the drill must be prevented from rotating in the chuck otherwise it cannot cut. This rotation is prevented by the friction between the chuck jaws and the drill shank. Feeding the drill into the work tends to cause the drill to push up into the chuck. This tendency is also prevented by friction between the chuck jaws and the drill shank. In order to supply sufficient friction the chuck must be well tightened.

A drill with tapered shank is shown in Fig. 4.84. The taper shank fits into a tapered hole in the drilling-machine

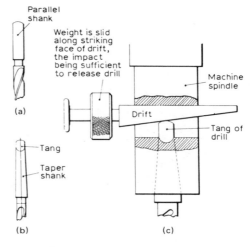

Parallel shank

Weight is slid along striking face of drift, the impact being sufficient to release drill

Machine spindle

(a)

Drift

Tang of drill

Tang

Taper shank

(b)   (c)

Fig. 4.84

spindle and is housed by a light tap with a soft hammer or by gently pressing the drill on to the work using the hand feed of the machine. The *tang* takes the force of the drift when removing the drill.

Fig. 4.84 shows the method of removing the drill from the machine spindle by means of a drift. A plain drift may also be used.

Sometimes the taper on the drill is not the same as the taper in the machine spindle. In this case a special taper socket is used as shown in Fig. 4.85. Fit the drill into the taper socket and then fit both into the machine spindle.

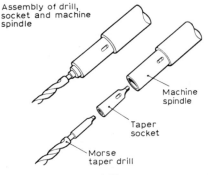

Assembly of drill, socket and machine spindle

Machine spindle

Taper socket

Morse taper drill

Fig. 4.85

When the drill is fitted into the tapered hole of the drilling-machine spindle the fit is so good that the drill can neither rotate in the tapered hole nor push up into the spindle. These effects are caused mainly by friction.

Holding the work. Do not hold the work in the hand because:
(1) it is dangerous to do so;
(2) this often causes drill breakage;
(3) inaccurate work may result.

Several methods of holding the work are shown in Fig. 4.86. Make sure that the drill does not cut the table or the machine vice after it has passed through the work.

All of the methods shown in Fig. 4.86 prevent the work from moving during drilling because of the restraints provided (see Chapter 4.3).

Safety
(1) A guard is required by <u>law</u> to cover the spindle and that part of the drill which is not cutting. On no account should the guard be moved—it is there to protect you.
(2) Never hold the work in your hand. It is very dangerous to do so.

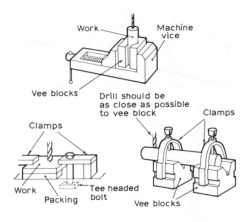

Work   Machine vice

Vee blocks   Drill should be as close as possible to vee block   Clamps

Clamps

Work   Tee headed bolt
Packing   Vee blocks

Fig. 4.86   Holding the work for drilling

(3) Make sure that the chuck key is removed before starting the machine.

Exercise 4.6
1. State the purposes of the following parts of a drill: (a) the shank; (b) the flutes; (c) the taper on the drill body.
2. Answer the following questions on drills:
   (a) What is body clearance?
   (b) Why is clearance provided at the point?
   (c) How is the rake angle provided?
3. When would a quick-helix drill be used?
4. What happens when a drill with unequal lip lengths is used?
5. Why is the web of a drill thinned?
6. What is meant by sensitive feed?
7. State three reasons which can account for drill breakage.
8. Give two reasons why the web of a drill may split.
9. Why are reamers used? Why will a reamer not correct a hole which has been incorrectly positioned?
10. Give two reasons for spotfacing a hole.
11. A cored hole in a casting is to be opened up. What kind of a drill must be used and why?
12. Why is it important to use: (a) the correct cutting speed; (b) the correct feed when drilling a hole?

4.7 Screw threads

The ISO metric screw thread. Bolts, nuts and screws are used for fastening parts together. The standard thread is the ISO screw thread (BS 3643), whose basic form is shown in Fig. 4.87.

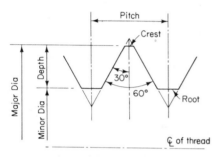

Fig. 4.87 Basic form of the ISO screw thread

The thread-form used in practice is slightly different (Fig. 4.88), mainly to avoid sharp corners at the crest and root.

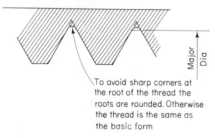

To avoid sharp corners at the root of the thread the roots are rounded. Otherwise the thread is the same as the basic form

Practical form of the ISO nut thread

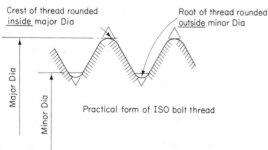

Crest of thread rounded inside major Dia

Root of thread rounded outside minor Dia

Practical form of ISO bolt thread

Fig. 4.88 Modifications to the basic form of the ISO screw thread

The *pitch* is the distance measured from a point on one thread to the same point on the next thread. When a bolt has a single continuous thread the pitch is the distance that the nut moves in one complete revolution. The pitch is usually given in millimetres.

A screw thread may have a fine pitch or a coarse pitch. When the pitch is fine the nut moves only a short distance for each turn of the screw, whilst with a coarse pitch the nut travels farther. Fine threads have a greater locking power than coarse threads and hence are used for screws which are subject to vibration.

Fig. 4.89 gives a comparison between a 10 mm diameter coarse thread and a 10 mm diameter fine thread. It will be seen that as well as having a smaller pitch the fine thread also has a smaller depth of thread.

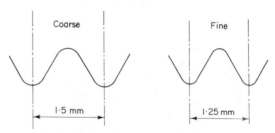

Fig. 4.89 Comparison between ISO coarse and fine threads

The table overleaf gives details of some of the threads which are available.

Coarse threads are indicated on drawings by the symbol M. Thus M 24 means a thread with a major diameter of 24 mm which is of the coarse type. From the table overleaf the pitch of this thread is 3 mm.

For fine threads the notation is M 24×2 which indicates a thread having a major diameter of 24 mm and a pitch of 2 mm. The letters RH and LH indicate right-hand and left-hand threads respectively. Thus M 42×3 LH indicates a fine left-hand thread having a major diameter of 42 mm and a pitch of 3 mm.

Cutting screw threads on the bench
1. *Internal threads*. Taps are used for cutting internal threads. The names of the various parts are shown in Fig. 4.90.

ISO METRIC SCREW THREADS (BS 3643)

Major diameter (mm)	Coarse threads			Fine threads		
	Minor diameter (mm)	Pitch (mm)	Tapping size (mm)	Minor diameter (mm)	Pitch (mm)	Tapping size (mm)
3	2·387	0·5	2·5	—	—	—
4	3·141	0·7	3·3	—	—	—
5	4·019	0·8	4·2	—	—	—
6	4·773	1	5	—	—	—
8	6·466	1·25	6·8	6·773	1	7
10	8·160	1·5	8·5	8·467	1·25	8·8
12	9·853	1·75	10·2	10·467	1·25	10·8
16	13·546	2	14	14·160	1·5	14·5
20	16·933	2·5	17·5	18·160	1·5	18·5
24	20·319	3	21	21·546	2	22
30	25·706	3·5	26·5	27·546	2	28
36	31·093	4	32	32·319	3	33
42	36·479	4·5	37·5	38·319	3	39
48	41·866	5	43	44·319	3	45

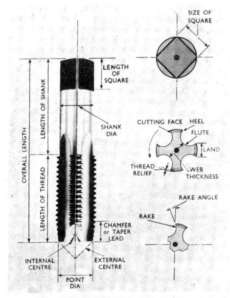

Fig. 4.90   The tap used for cutting internal threads

Taps are usually supplied in sets of three to cut a particular size of thread.

All three taps have the same maximum diameter.

The *taper tap*, Fig. 4.91 (a), is used first, because the taper allows the tap to cut the full length gradually.

The *second tap*, Fig. 4.91 (b), is used for finishing threads when the hole is open at both ends.

The *plug tap*, Fig. 4.91 (c), is only used when the thread is in a blind hole or if the thread has to be taken up to a shoulder.

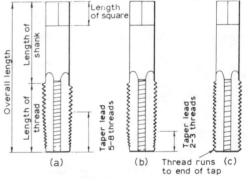

Fig. 4.91   (a) Taper tap
            (b) Second tap
            (c) Plug or bottoming tap

Tap wrenches, Fig. 4.92, are used to form a handle for the tap.

2. *Drilling and tapping the hole*. Before the thread can be cut a hole must be drilled in the work. The size of the hole must be such that there is enough metal left to form the thread. This hole should be slightly larger than the minor diameter of the thread, because the thread is then easier to cut. The size of hole which is drilled ready for tapping is called the *tapping* size. Tapping sizes are usually given in screw thread tables.

3. *External threads*. For cutting external threads stocks and dies are used, Fig. 4.93.

Square and buttress threads. Square threads (Fig. 4.94) are used for moving parts because there is less friction with a square thread than the vee type.

Fig. 4.94

The square thread

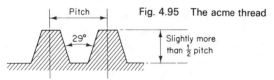

The square thread is very difficult to cut by using taps and dies and for this reason an *acme* thread (Fig. 4.95) is often used instead.

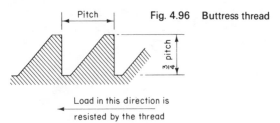

Fig. 4.95   The acme thread

The *buttress* thread (Fig. 4.96) is designed to resist heavy axial loads in one direction. Vices and similar equipment are usually fitted with buttress threads.

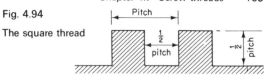

Fig. 4.96   Buttress thread

Load in this direction is
resisted by the thread

Exercise 4.7

1. By means of sketches show what is meant by:
   (a) the pitch; (b) the depth of a screw thread.
2. Using clear sketches show the difference between the basic form of the ISO thread and the practical form of the thread.
3. Draw ten times full size the thread form of a 20 mm diameter coarse thread.
4. When is a fine thread used in preference to a coarse thread?
5. Using the tape of ISO screw threads (page 154) state: (a) the pitch (b) major diameter (c) minor diameter of a screw thread marked M 30×2 on a drawing.
6. Taps are usually supplied in sets of three. What is each tap in the set called? What are the main differences between these taps? State the purpose for which each tap in the set is used.
7. By means of a clear sketch show how rake and clearance are provided on a tap.
8. Why is the tapping-size of a hole made slightly larger than the minor diameter of the thread? What effect does this have on the thread?
9. For what purpose is a die nut used?
10. Sketch (a) a square thread (b) an acme thread (c) a buttress thread. Why is an acme thread often used instead of a square thread? When would a buttress thread be used?

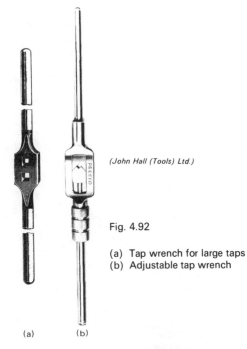

(John Hall (Tools) Ltd.)

Fig. 4.92

(a) Tap wrench for large taps
(b) Adjustable tap wrench

(a)   (b)

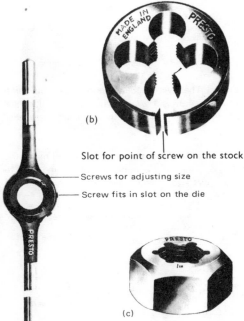

(b)

Slot for point of screw on the stock

Screws for adjusting size

Screw fits in slot on the die

(c)

(a)

Fig. 4.93   Stocks and dies:
(a) Stock for split die. Screw fits in slot on the die;
(b) Split die for small threads;
(c) Die nut used for cleaning up threads. It is not used for cutting threads but only for bringing oversized threads to size

4.8   Assembly work

Riveted joints. Fig. 4.97 shows some typical riveted joints and the proportions that should be used if the joint is to have maximum strength.

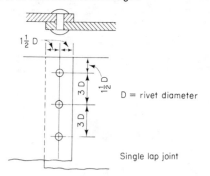

D = rivet diameter

Single lap joint

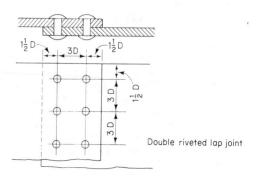

Double riveted lap joint

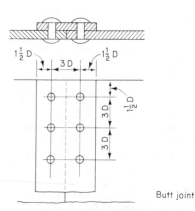

Butt joint

Fig. 4.97   Types of riveted joints

Types of rivets. Fig. 4.98 shows various types of rivets and their uses. Riveted joints are made (wherever possible) so that the rivets are in shear. It is generally bad practice to allow rivets to be in tension.

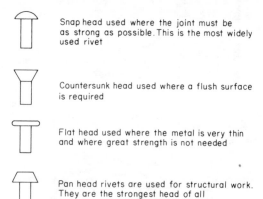

Snap head used where the joint must be as strong as possible. This is the most widely used rivet

Countersunk head used where a flush surface is required

Flat head used where the metal is very thin and where great strength is not needed

Pan head rivets are used for structural work. They are the strongest head of all

Fig. 4.98   Types and uses of rivets

Rivet size. The rivet diameter required for the joints shown in Fig. 4.97 is $1\cdot2\times\sqrt{t}$, where t is the thickness of the plates. For thin plates this formula is often amended to $1\cdot4\times\sqrt{t}$.

Defects in riveted joints. Holes for riveting may either be drilled or punched, the punched holes being used for the larger rivet sizes. If the holes are to be drilled they should be clamped together and then drilled, which

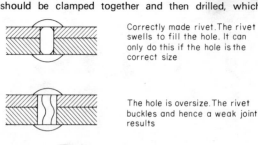

Correctly made rivet. The rivet swells to fill the hole. It can only do this if the hole is the correct size

The hole is oversize. The rivet buckles and hence a weak joint results

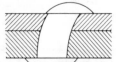

The holes do not line up properly and the head has been badly made

Fig. 4.99   Defects in riveted joints

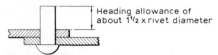

Heading allowance of about 1½ × rivet diameter

Fig. 4.100 The heading allowance for a rivet

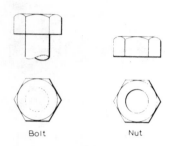

Bolt   Nut

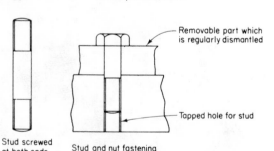

Removable part such as a casing

Hexagon set screw which is screwed to the head

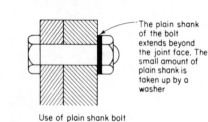

The plain shank of the bolt extends beyond the joint face. The small amount of plain shank is taken up by a washer

Plain shank bolt   Use of plain shank bolt

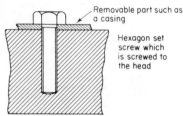

Removable part which is regularly dismantled

Tapped hole for stud

Stud screwed at both ends   Stud and nut fastening

Countersunk head   Round head   Cheese head

Fig. 4.101 Screwed fastenings

makes sure that they line up properly. When the plates are drilled separately or punched, the plates may be strained after riveting because the holes do not match up properly.

It is essential that the rivets fit properly in the holes. The idea is that the rivet shank should completely fill the hole (Fig. 4.99).

The length of the rivet is also important. If too short a rivet is used a weak head will be formed. Too long a rivet will result in the rivet buckling, again causing a bad head to be formed. Fig. 4.100 shows the length of rivet that should be used.

Screwed fastenings. These are used where parts need to be taken apart and reassembled. Various kinds of fastenings are shown in Fig. 4.101.

Locking of bolts and screws. When a nut is screwed on a bolt there is always the danger that the nut will unscrew due to vibration, etc. Many nuts are therefore locked on the bolt. The methods of locking may be positive or fractional.

Positive Locking. Fig. 4.102 illustrates:

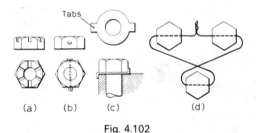

Tabs

(a)   (b)   (c)   (d)

Fig. 4.102

(a) Slotted or castle nut. A split pin is used to lock the nut, the legs of the pin being bent back round the cylindrical portion of the nut.

(b) Split pin and ordinary hexagon nut. A hole is drilled in the nut and bolt and a split pin is inserted as shown.

(c) Tab washers. The washer is placed on the bolt and the nut is tightened down, the tabs on the washer are

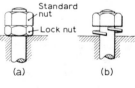

Standard nut

Lock nut

(a)   (b)

Fibre or plastic insert

(c)

Fig. 4.103

then turned up against the flat of the bolt and the edge of the component as shown.

(d) Locking wire. The head of each screw is drilled and soft iron or copper wire is drawn through the holes and pulled tight, and tied.

Frictional Locking. Fig. 4.103 illustrates:

(a) Lock nuts. The lock nut is screwed tightly down on the bolt and the standard nut is then screwed tightly on to the lock nut. The lock nut is then *screwed back* on to the standard nut which is held by a spanner. The threads of the two nuts become wedged between the threads of the bolt. The lock nut is about half the thickness of the standard nut.

(b) Spring washer. The spring washer is placed under the nut and the nut screwed down. The washer exerts an upward force on the nut and this increases the friction between the threads of the bolt and nut. Both single-coiled and double-coiled washers can be obtained.

(c) Fibre insert lock nut. The bolt cuts its own thread in the fibre or plastic. The nut is held tight by friction between the bolt and the fibre.

Assembly of bolted and screwed joints

(1) Wherever possible a bolted joint should be prepared by marking out the holes in the top plate, clamping the plates together and drilling all the plates together. This method ensures that the holes line up.

(2) When screws are used for assembling, a clearance hole is needed for the bolt in one plate, whilst a hole for tapping is needed in the other, Fig. 4.104. The top plate is marked out for drilling the holes and the plates then clamped together. Both plates are then drilled to the *tapping size* and taken apart. The holes in the top plate can then be opened out and the bottom plate tapped.

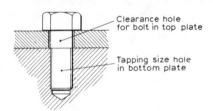

Fig. 4.104   Plates assembled by using screws

Relationship between surfaces. Before we can fit mating parts together we have to check the accuracy of their surfaces to see if they possess flat surfaces, if certain edges are square, etc.

1. *Check for flatness.* We check for the flatness of a surface by comparing it with a known flat surface. The surface plate has a surface of proven flatness and hence we compare the surface of the job with the surface of the plate. To do this, the surface plate is smeared with blue. The surface to be tested is then placed in contact with the surface plate and moved about. If the job is reason-

ably flat, spots of blue will be visible all over the tested surface.

A *straight edge* can be used by placing it on the job and looking for 'daylight'. Flatness can be judged to about 0·005 mm by this method.

Fig. 4.105   Straight edge. Lengths available from 300 mm to 5000 mm made from cast iron

2. *Squareness* can be checked with a try square or the combination square.
3. *Parallelism* can be checked by using calipers, micrometers, height gauges, scribing blocks, etc. Some methods of testing parallelism by using a scribing block are shown in Fig. 4.106.

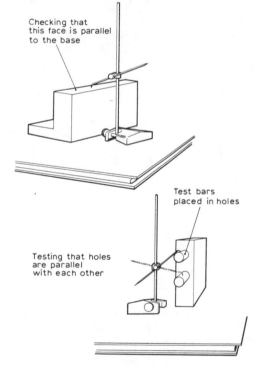

Fig. 4.106

The scribing block method of checking parallelism has the following disadvantages:

(a) The scribing block does not give a measure of any errors in parallelism.

(b) The accuracy of the test depends upon the 'feel' of the operator.

These disadvantages are overcome by using a dial gauge, Fig. 4.107.

Slight pressure on the plunger causes the needle to move. This type relies on a rack and pinion followed by a gear train to magnify the plunger movement.

Fig. 4.107   Dial-gauge set

4. *Concentricity and roundness.* With some round work it is important that the various diameters have the same centre. This is known as *concentricity* and is best checked by using a dial indicator. The method is described in Fig. 4.108. Roundness can also be

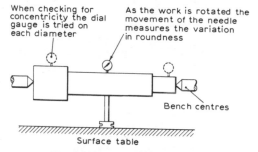

When checking for concentricity the dial gauge is tried on each diameter

As the work is rotated the movement of the needle measures the variation in roundness

Bench centres

Surface table

Fig. 4.108   Check for concentricity

checked with a dial indicator. The work may be supported on bench centres or in a vee block, Fig. 4.109.

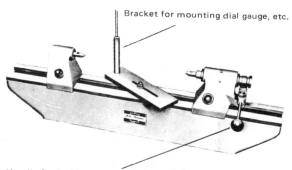

Bracket for mounting dial gauge, etc.

Handle for locking centre after it has been positioned

Fig. 4.109   Bench centres

Exercise 4.8

1. State where the following kinds of rivets would be used: (a) snap head; (b) countersunk head; (c) flat head.
2. Sketch a double-riveted lap-joint showing the pitch of the rivets and their edge distances. The rivets are to be 12 mm diameter.
3. Two plates which are 16 mm thick are to be riveted together. What rivet size should be used?
4. Show two defects that can occur in riveted joints stating the reason why each defect occurs.
5. Why is the length of a rivet important? Find the length of rivet needed if a 6 mm diameter rivet is used for riveting two plates 25 mm thick together.
6. State two reasons why a bolted joint may be preferred to a riveted joint.
7. Make sketches of: (a) a stud; (b) a bolt; (c) a hexagon-headed set-screw.
8. When would (a) countersunk-head screws (b) cheese-head screws be used?
9. Why are nuts locked on a bolt? Describe, with sketches, two methods of positive locking and two methods of frictional locking.
10. How does a spring washer assist in locking a nut?
11. Why should the holes for bolts be drilled in all the parts simultaneously?
12. What procedure should be adopted when making an assembly where screws are to be used?
13. Give two methods by which flatness can be checked.
14. What are the advantages of a dial gauge over a scribing block when checking for parallelism?
15. What is concentricity? How may it be checked?
16. Give two methods of checking for roundness.

4.9   Forging

Forging is the process of shaping metals by hammering them. To do this the metals are made plastic by heating them and the shaping process takes place whilst they are hot. Forging is the craft of the blacksmith and it is the oldest method of shaping metal.

The forge is needed for heating metal.

(a)

(Wm. Allday & Co. Ltd.)

Extractor pipe for removing fumes from the fire

Tank containing water for cooling tuyere. Tuyere can then stick well out into the fire without its nose getting burnt

Hearth which holds the fuel. The hearth is lined with firebrick

Tuyere through which draught of air reaches the fire. Temperature of fire depends on the amount of air blast supplied. Air blast is supplied by the blower

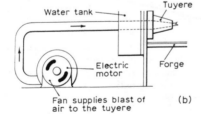

Water tank

Tuyere

Electric motor

Forge

Fan supplies blast of air to the tuyere

(b)

Fig. 4.110   (a) Blacksmith's forge
(b) Illustrating how the air blast reaches the tuyere

The fire. The best fuel is coke breeze, which is coke crushed into small pieces about 12 mm diameter. The shape of the fire can be arranged to suit the job in hand: for example, a long narrow fire will suit a long piece of metal. The amount of air draught is important—too little causes the fire to die out whilst too much causes the fire to blow out in a shower of sparks. A covering of dead coke raked over the top of the fire will act as an insulating layer to keep the heat of the fire down on the metal. This will also make the fire heavy and less likely to blow out.

The anvil is used for supporting the work whilst it is being hammered.

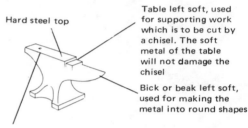

Hard steel top

Table left soft, used for supporting work which is to be cut by a chisel. The soft metal of the table will not damage the chisel

Bick or beak left soft, used for making the metal into round shapes

Hardy hole for supporting various tools

Fig. 4.111   Blacksmith's anvil

Forging tools. Fig. 4.112 illustrates two types of hammer used in forging.

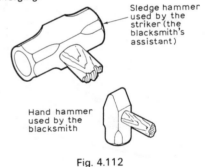

Sledge hammer used by the striker (the blacksmith's assistant)

Hand hammer used by the blacksmith

Fig. 4.112

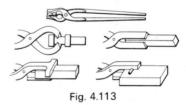

Fig. 4.113

Fig. 4.113 shows tongs used for holding the work and cutting on the anvil. Various patterns of tongs are used according to the shape of the work to be held.

Figs 4.114 and 4.115 illustrate chisels and hardies used for cutting on the anvil.

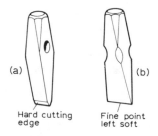

(a)   (b)

Hard cutting   Fine point
edge   left soft

Fig. 4.114 (a) Cold set used for cutting cold metal
(b) Hot set used for cutting hot metal.
Quenching is necessary after three or four
blows

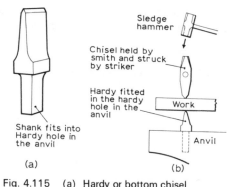

Sledge
hammer

Chisel held by
smith and struck
by striker

Hardy fitted
in the hardy
hole in the   Work
anvil

Shank fits into   Anvil
Hardy hole in
the anvil

(a)   (b)

Fig. 4.115 (a) Hardy or bottom chisel
(b) Using a hardy and chisel together

Punches may be round, square or any other shape as
required. They are used for punching holes in hot metal.
For deep holes the punch should be removed after two or
three blows and quenched in water. Punched holes are
superior to drilled holes because the displaced metal
bulges round the hole, preserving strength.

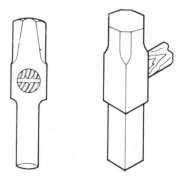

Round punch   Square punch
Fig. 4.116

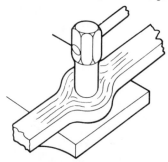

Fig. 4.117 Punching a hole. For deep holes the punch
should be removed after two or three blows
and quenched in water

Drifts are used for opening out punched holes. They are
driven right through the punched hole. They shape and
smooth the hole as well as enlarging it.

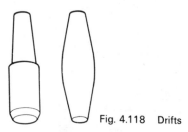

Fig. 4.118   Drifts

Fullers are described in Figs 4.119 and 4.120.

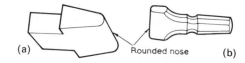

(a)   Rounded nose   (b)

Fig. 4.119 Fullers used for reducing a bar:
(a) Bottom fuller. Shank fits in hardy hole;
(b) Top fuller. Held by the smith and struck
by the striker

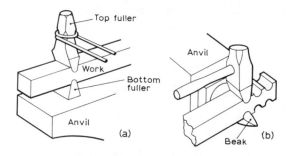

Top fuller

Anvil

Work

Bottom
fuller

Anvil

(a)   Beak   (b)

Fig. 4.120 (a) Fullers used to neck the bar before
reducing it
(b) Drawing down a bar using a fuller and
the beak of the anvil

A flatter is used for finishing flat work. The tool is placed on the work by the smith and it is struck by the striker using a sledge hammer.

Fig. 4.121   Flatter

Swages can be obtained for working the metal to any shape. Those illustrated are for round work.

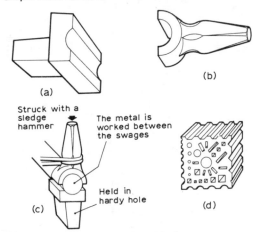

(a)

(b)

Struck with a sledge hammer

The metal is worked between the swages

Held in hardy hole

(c)

(d)

Fig. 4.122   Swages and swage block:
(a) Bottom swage. Shank is fitted in the hardy hole;
(b) Top swage. Struck with a sledge hammer;
(c) Method of working;
(d) Swage block containing many different shapes and sizes of swage

Forging operations

1. *Drawing down* is the process of increasing the length of the bar by reducing its width or thickness or both.

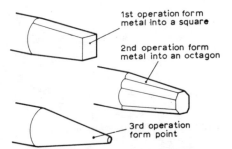

1st operation form metal into a square

2nd operation form metal into an octagon

3rd operation form point

Fig. 4.123   Operations for forming a point on a round bar

Methods of drawing down have been previously illustrated. The method of forming a point on a round bar is shown in Fig. 4.123.

2. *Bending* can be carried out cold but it is best done when the metal is hot. Methods of bending are illustrated in Fig. 4.124.

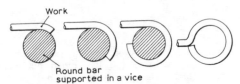

Work

Round bar supported in a vice

Fig. 4.124   Operations for forming an eye

3. *Upsetting or jumping up* consists of increasing the cross-section of the work at one particular place. By doing this the length of the work is shortened.

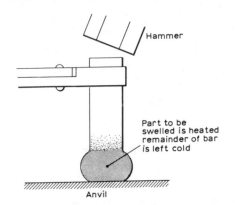

Hammer

Part to be swelled is heated remainder of bar is left cold

Anvil

Fig. 4.125   Upsetting

FORGING TEMPERATURES FOR STEEL AND WROUGHT IRON

Colour	Temp. °C	Uses
Dull red	700	Easy bends on mild steel.
Bright red	900	Simple forging on mild steel. Light bending and chiseling.
Bright yellow	1100	For all principal forging operations on steel and wrought iron.

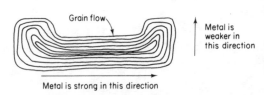

Grain flow

Metal is weaker in this direction

Metal is strong in this direction

Fig. 4.126   Grain flow as a result of forging

Materials used in forging

Metal	Forging temp. range °C	Remarks
Wrought iron	860–1340	Easily forged since it possesses ductility and malleability. Largely superseded by mild steel.
Mild steel	820–1290	Readily forged. Stronger than wrought iron, and cheaper.
Medium-carbon steel	760–1250	Reasonably easy to forge but not so easy as mild steel. Harder and stronger than mild steel.
High-carbon steel	760–1130	More difficult to forge than any of the preceding metals. Because of small range of forging temperatures may easily be burnt and spoiled. Used for cutting tools.
Aluminium alloy	340–450	Wrought alloys are very suitable for forging. Used in the aircraft industry.
Brass (60–40)	600–800	Readily forged.

Non-forging materials. Some metals are apt to break and crumble when hammered at red heat. Such metals are called 'red short' and are unsuitable for forging. *Cast iron* is such a metal and cannot be used for any forging operation.

The effect of forging. When steel is forged the original crystal structure is deformed. In effect the crystals are

drawn out into threads and the resemblance to the grain of wood is most marked (Fig. 4.126). The forged metal is very strong in the direction of the grain flow, but it is weaker in a direction at right-angles to the grain flow.

Aluminium alloys are forged at 340 °C–450 °C. These temperatures are much lower than those used for forging ferrous metals. If aluminium alloys are overheated they crack easily when forged. The metal must be rapidly heated to the forging temperature because slow heating will give a coarse grain structure. Fairly light blows must be used because heavy blows cause surface cracks which cannot be welded up by further forging, as is the case with ferrous metals. The edges of fullers and swages must be well rounded to avoid nicks appearing on the work.

The effect on brass is similar to that for aluminium.

Exercise 4.9

1. For what purpose is the tuyere used?
2. Why is the tuyere water-cooled?
3. Why is the table of the anvil left soft?
4. What is the hardy hole used for?
5. What is the main difference between hot and cold sets?
6. For what purposes are the following tools used? (a) cold set, (b) hardy, (c) drifts, (d) hot set, (e) fullers, (f) swages, (g) flattener.
7. Name two metals that are suitable for hand forging and one that is unsuitable.
8. Why is a punched hole to be preferred to a drilled hole?
9. Describe the operation known as upsetting.
10. Describe the operation of drawing-down stating the tools that would be used and the heat colour needed.
11. Arrange the following metals in order of their suitability for forging (give the best first): (a) cast iron; (b) mild steel; (c) carbon tool steel; (d) wrought iron.
12. When is a metal said to be 'red short'? Give an example of such a metal.
13. What is the effect on the crystal structure of steel when the metal is forged?
14. What difficulties occur when forging aluminium and its alloys?

4.10　Fabrication processes

Shears

Note the following points:

(1) The size of the shears depends on the thickness of the metal being cut. Shears with longer handles should be used when thicker sheet is being cut, as more leverage is then obtained.

(2) When cutting tougher metals use shears with larger blade sizes to prevent the blades from being strained.

(3) Before using shears make sure that the blades are sharp. The blades are sharpened along the edges only and never on the faces.

(4) Do not try to cut hardened steel with shears.

(5) Keep the blades clean and the pin lubricated.

(6) Always take small cuts. On small work move the work back towards the pin. On large work move the shears forward and start each cut as near the pin as possible. Do not allow the shears to close completely on one cut or a ragged edge will result.

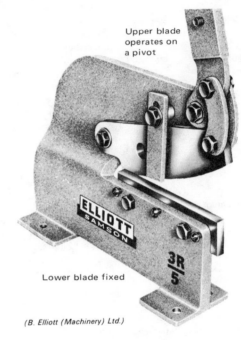

Upper blade operates on a pivot

Lower blade fixed

(B. Elliott (Machinery) Ltd.)

Fig. 4.128　Hand shearing machine

(F. J. Edwards Ltd.)

Fig. 4.127　(a) Straight-type shears. Used for straight cuts and for trimming off surplus metal
(b) Curved-type shears. Used when curves have to be cut. One handle is set well above the blade pivot to give free movement on the surface being cut

Hand shearing machine (cropper). This is used for cutting thick metal. The machine illustrated in Fig. 4.128 will cut 3 mm sheet and 5 mm strip. As with hand shears do not use the full length of the blade or a ragged edge will result.

The *guillotine* is the most widely-used of the straight-line cutting-machines. Its cutting action is very similar to that of the hand shearing machine. The guard, which is attached between the operator and the cutting blades should be set in such a way as to protect the hands from the blades.

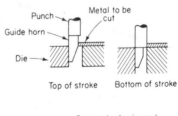

Punch

Metal to be cut

Guide horn

Die

Top of stroke　　Bottom of stroke

Serrated edge in work

Fig. 4.129　The nibbling machine

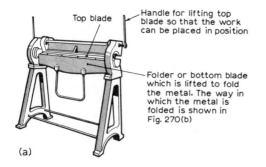

Top blade

Handle for lifting top blade so that the work can be placed in position

Folder or bottom blade which is lifted to fold the metal. The way in which the metal is folded is shown in Fig. 270(b)

(a)

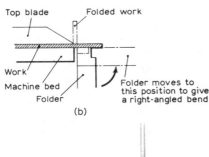

Top blade   Folded work

Work

Machine bed

Folder

Folder moves to this position to give a right-angled bend

(b)

(c)

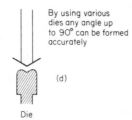

By using various dies any angle up to 90° can be formed accurately

(d)

Die

Fig. 4.130   (a) Cramp folder   (b) Method of folding
               (c) Press brake   (d) Method of forming

The nibbling machine. The main uses of this machine are:

(1) for cutting curved shapes;

(2) for cutting narrow slots;

(3) for cutting holes which are some distance from the edge of the sheet. A hole must be drilled or punched to allow the nibbling punch to enter the sheet-metal blank before the final adjustment of the machine.

The principle of the machine is shown in Fig. 4.129. A small piece of metal is removed with each cutting stroke. The action is unlike that of shears which produce a cut of negligible width. The width of the cut on the nibbling machine depends upon the size of the punch used.

Bending and folding machines are used for bending sheet metal.

Both the cramp folder and the press brake are used for bending and folding metal. The cramp folder works in a rotary manner as shown in Fig. 4.130 (b) but the press brake forms the metal by the action of the top beam pressing the metal into the die (Fig. 4.130 (d)).

To produce a clean bend the blade over which the metal is bent must be straight, smooth and reasonably sharp. The pressure, which is applied to make the bend, must be the same along the complete length of the bend.

Stakes. These are used for supporting the work whilst it is hammered to shape. Many types are used according to the shape of the article and the resistance offered by the metal.

Stakes. Fig. 4.131 illustrates:

(a) Hatchet stake. Used for cleaning up folded work and for closing down edges:

(b) Half-moon stake. Used for similar operations to the hatchet stake, but on circular work.

(c) Bottom stake. Used for circular work and on round tops and bottoms.

(d) Round head and ball head stakes. Used for shaping articles of a curved shape.

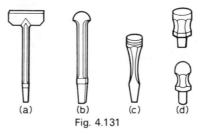

(a)   (b)   (c)   (d)

Fig. 4.131

Rolling is done to produce a curved shape.

Bend allowance. When a fold is made the metal at the outside of the bend stretches whilst the metal shortens at the inside of the bend. When accurate work is needed an allowance must be made for this effect. The bend

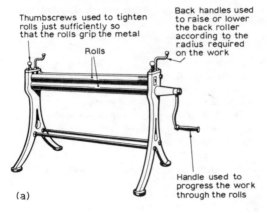

Thumbscrews used to tighten rolls just sufficiently so that the rolls grip the metal

Rolls

Back handles used to raise or lower the back roller according to the radius required on the work

Handle used to progress the work through the rolls

(a)

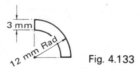

Front roll

Back roll is free to revolve

Metal should always be rolled downwards as shown

Two front rolls geared together

Metal is progressed in this direction by turning the handle

Fig. 4.132   (a) Rolling machine   (b) Method of rolling

allowance is calculated by finding the average radius between the inner and outer radii of the bend and then working out the length of arc.

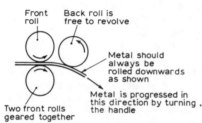

3 mm

12 mm Rad

Fig. 4.133

In Fig. 4.133 outer radius = 12 mm
            inner radius = 12−3 = 9 mm
          average radius =10·5 mm

To calculate the bend allowance we now have to find the circumference of a quarter-circle whose radius is 10·5 mm.

Bend allowance = $\frac{1}{4} \times 2 \times \pi \times 10\cdot5$ = 16·5 mm

Example. Find the width of strip needed to make the section shown in Fig. 4.134 (a).

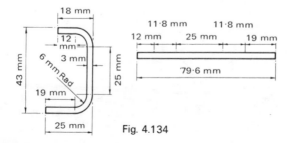

18 mm

12 mm

3 mm

6 mm Rad

43 mm

19 mm

25 mm

25 mm

11·8 mm   11·8 mm
12 mm   25 mm   19 mm

79·6 mm

25 mm

Fig. 4.134

The easiest way of making this section is to use a bend block.

Average radius at corners = 7·5 mm.

Bend allowance at each corner = $\frac{1}{4} \times 2 \times \pi \times 7\cdot5$ = 11·8 mm.

The width of metal required is shown in Fig. 4.134 (b).

Joints in sheet-metal work

Grooved joint

Width of joint

Locked groove joint

Fig. 4.135

The grooved joint is easy to make, but the locked groove joint is stronger. Both these joints are airtight and water-tight.

The bend allowance for the locked groove joint can be calculated from the formula:

bend allowance = 3 times width of groove−2 times thickness of material

Beading and wiring. The edge of sheet metal is weak and sometimes sharp enough to cut. Beading and wiring is done to strengthen the edge and to make it safe. A wired edge is the stronger. A paning hammer is used for this work.

Wire inserted under beading

Beaded edge   Wired edge

Fig. 4.136

To form a beaded edge the metal is first formed on a wire and then the wire is removed. The bending allow-ance for both wiring and beading can be calculated from:

bending allowance = $2\frac{1}{2}$ times the wire diameter

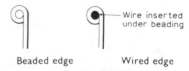

Fig. 4.137   Paning hammer

Stiffening of sheet metal. When sheet metal is used in large panels it tends to buckle because of its lack of strength during bending. Much of this buckling may be prevented by using stiffeners to reduce the size of the unsupported metal. Angle formers are often used as stiffeners but other structural shapes are used. Fig. 4.138 shows an example of a stiffened panel.

Punching holes. It is often undesirable to drill holes in thin plate and so the holes are frequently punched out. A drilled hole will often be ragged and of poor shape but

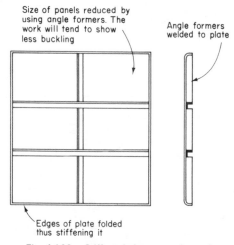

Fig. 4.138   Stiffened sheet-metal panel

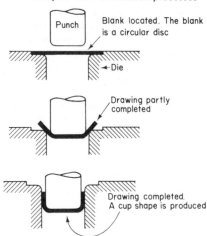

Fig. 4.140   Drawing to produce a cup shape from a flat disc

a punched hole will have a good finish and shape. Fig. 4.139 shows the way in which a hole is punched. Presses ranging from the small hand-operated fly-press to the heavy machine-operated presses are used. To make sure that the hole is punched in the correct place the work is located on the die often by means of dowels.

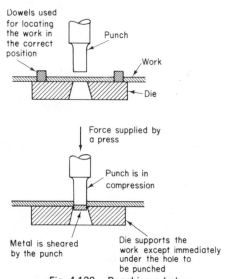

Fig. 4.139   Punching a hole

Drawing. When a cup or a dish shape is needed these are usually produced by drawing. The process is illustrated in Fig. 4.140.

Spinning. When hollow shapes with circular cross-sections are required these are often produced by spinning. It is cheaper than pressing when a small number of parts are needed. A high-speed lathe and a wooden former

are essential for the spinning operation which is illustrated in Fig. 4.141.

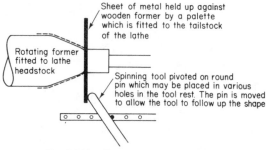

Fig. 4.141   The operation of spinning

Safety
(1) Shearing machines and guillotines are very dangerous. Never use shearing machines unless the guards are in position.
(2) The edges of sheared plates are sharp. Always use gloves when handling them.
(3) The edge of sheet-metal parts are made safe by folding, beading or wiring.

Exercise 4.10
1. State when the following tools would be used: (a) straight shears; (b) curved shears; (c) cropper.
2. When would a hole cutter be used?
3. What determines the size of shears that would be used for a particular job?
4. The blade of a folding machine must possess three essential features if a clean bend is to be produced. What are they? What other factor is essential for a good bend?

5. Sketch three kinds of stake and state their uses.
6. State three metals used in sheet-metal work.
7. By means of a clear diagram show how sheet metal is rolled.
8. What is a wired edge? Why is it used?
9. Find the length of sheet metal needed to make the shapes shown in Fig. 4.142.
10. Why are panels made from thin sheet stiffened?
11. By means of a clear sketch show how a hole is punched?
12. Illustrate the difference between drawing and spinning.

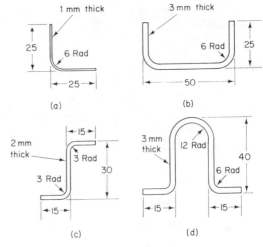

All dimensions in millimetres

Fig. 4.142

4.11   Soldering and brazing

Soldering is a quick and useful method of joining metals. It is used to make joints in light articles made from copper, brass, steel, etc. Solder has a low melting temperature (see table below) and it must not be used where the joint has to withstand heat. A soldered joint is not very strong and if much strength is needed other methods of making the joint, such as riveting, brazing or welding, must be used.

Soft solders. Soft solders are lead-tin alloys and some also contain a small amount of antimony. The table below gives the most widely used of these.

SOFT SOLDERS

Tin %	Lead %	Anti-mony %	Melting range °C	Applications
65	34·4	0·6	183–185	Electricians' solder used for soldering electrical components which may be damaged by heat.
60	39·5	0·5	183–188	Tinman's solder. Quick-setting solder for good class sheet-metal work.
50	47·5	2·5	183–204	Tinman's solder for general workshop use. Cheaper than the above.
30	69·7	0·3	183–255	Plumbers' solder. The high lead-content causes the solder to remain pasty for a long enough period for the joint to be wiped.

Fluxes. For soldering to be successful the metal surfaces to be soldered must be absolutely clean. This is done by hand with a file, sandpaper or emery cloth. Large areas are usually pickled in an acid bath.

Fluxes help in maintaining the cleanliness. They also protect the cleaned surfaces from fumes and from chemical action due to the atmosphere. There are two common kinds of fluxes:
(1) those which merely protect the cleaned surfaces,
(2) those which protect the surfaces and also, by chemical action, help to clean it.

Resin, tallow, Vaseline, pure turpentine, etc., fall into the first group of fluxes. Zinc chloride (killed spirits), ammonium chloride and zinc ammonium chloride are fluxes which fall into the second group. The second group is the most useful and hence the most widely used.

The table below shows the fluxes which are used for various metals.

Metal	Flux
Steel	Ammonium chloride
Lead	Tallow
Zinc and galvanized work	Dilute hydrochloric-acid
Brass	Zinc chloride (killed spirits) or resin
Tin plate	Zinc chloride

Method of soldering. The soldered joint is made by heating the joint, applying the flux and adding the solder. Generally, the solder is applied by using a *soldering iron.* Fig. 4.143 illustrates the soldering process.

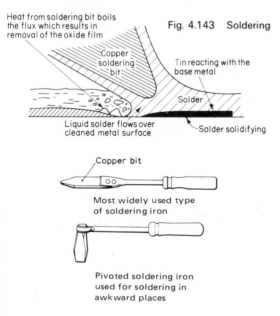

Heat from soldering bit boils the flux which results in removal of the oxide film

Fig. 4.143   Soldering

Copper soldering bit

Tin reacting with the base metal

Solder

Liquid solder flows over cleaned metal surface

Solder solidifying

Copper bit

Most widely used type of soldering iron

Pivoted soldering iron used for soldering in awkward places

The bit is made of copper because:
(1) this metal is a good conductor of heat and rapidly transfers heat from itself to the joint,

(2) it easily alloys with tin, thus allowing the end of the iron to be coated with solder. This is known as 'tinning' the iron.

The iron is usually heated in a gas heater.

Types of soldered joints

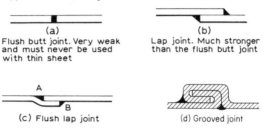

(a)	(b)

Flush butt joint. Very weak and must never be used with thin sheet

Lap joint. Much stronger than the flush butt joint

A

B

(c) Flush lap joint

(d) Grooved joint

Fig. 4.145   Soldered joints

Sweating. When really strong joints are needed they must be sweated. The flush lap joint, Fig. 4.145 (c), can be sweated by applying enough solder at A for it to run under the top sheet and emerge at B. Sufficient heat applied for a long enough time is essential. The joint surfaces must be tinned before sweating and when the heat is applied the tinned surfaces become molten and fuse together. The two surfaces in contact should be pressed together and sufficient time should elapse for the work to cool before it is moved. Any surplus solder can be removed by heating the work and wiping the joint with a cloth which has been dipped in flux.

Hard soldering and brazing. Joints made with soft solder are strong enough for most sheet-metal work purposes. When really strong soldered joints are needed hard solder must be used.

Brazing solder contains approximately equal parts of zinc and copper and it melts at about 870 °C. The metals to be joined have to be heated to a cherry-red colour in the neighbourhood of the joint. Joints in steel and brass are often brazed.

When silver is added to the brazing solder the melting-point of the solder is lowered. Such solders are called *silver solders* and they can be obtained with a melting-point as low as 630 °C. *Soft silver solders* do not contain either zinc or copper and they have a melting-point about 320 °C.

Several types of brazed joints are shown in Fig. 4.146.

The heat source used when brazing is a blowlamp or blowpipe. Sometimes in production brazing a furnace is used. The braze must be cooled in still air. Quick cooling reduces the strength of the parent metals but it does not effect the braze itself. Borax and water is used as a brazing flux.

Edges of plates
chamfered as shown
Vee joint

Brazing without vee
preparation

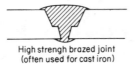

High strengh brazed joint
(often used for cast iron)

Fig. 4.146   Brazed joints

Exercise 4.11

1. When must soldering not be used for making a joint?
2. Why are fluxes used when soldering?
3. Name the fluxes recommended for soldering: (a) brass; (b) galvanized work; (c) steel.
4. Why is a soldering iron made from copper?
5. Sketch three types of soldered joint in common use.
6. What is the difference between soldering and sweating?
7. In making a certain joint it is essential that the solder should set quickly. State the composition of a suitable solder.
8. When is hard solder used in preference to soft solder?
9. What is the difference between brazing solder and silver solder? When would silver solder be used in preference to brazing solder?
10. Sketch a brazed joint suitable for cast iron.
11. Name two heat sources used when brazing.
12. Why must a brazed joint be cooled slowly. Why is the braze itself not affected by quick cooling?

4.12   Welding

Welding is used when a permanent joint which has high strength is required. It is often better to fabricate steel by welding than use a casting, because the mechanical properties of steel are better than those of cast iron. For many applications welding is superior to riveting and it is often cheaper and easier.

Fusion welding. When soldering and brazing (Chapter 4.11), the joints are formed by a thin film of metal which has a lower melting-point and less strength than the metals being joined. In fusion welding the edges of the metals to be joined are melted. The molten metal runs together and so a joint is made whose strength is equal to that of the metal being joined.

Oxy-acetylene welding. In this type of welding the heat source is provided by a flame of acetylene burning in an atmosphere of pure oxygen.

By varying the amounts of oxygen and acetylene it is possible to obtain one of the three types of flame shown in Fig. 4.147.

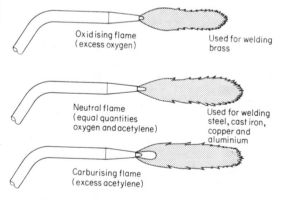

Fig. 4.147   Types of oxy-acetylene flames

The hottest part of the flame occurs just outside the luminous cone where a temperature of 3200 °C may be obtained.

Methods of welding

1. *Leftward welding* which is used for welding steel plate up to 3 mm thick and for welding non-ferrous metals. The method is shown in Fig. 4.148.

The welding rod is kept in front of the flame and it is moved in a straight line. The flame is not used to melt the rod – heat from the molten pool does this

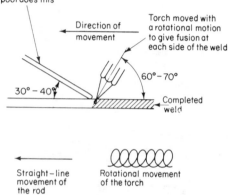

Fig. 4.148   Leftward welding

2. *Rightward welding.* It is difficult to weld thick steel plates by using the leftward method and the rightward method is then used (Fig. 4.149). The advantages of rightward welding are as follows:

(a) It is faster than leftward welding and less welding rod is used.

(b) There is a better view of the molten pool which means that there is better control over the weld.

(c) The blowpipe is moved in a straight line and hence there is less agitation of the molten metal so giving less oxidation.

(d) The flame plays on the weld which has just been completed thereby helping to anneal to the weld and thus making it less brittle.

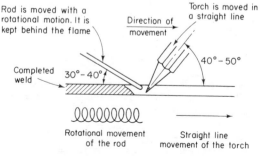

Fig. 4.149   Rightward welding

Welded joints. Fig. 4.150 shows some typical welded joints.

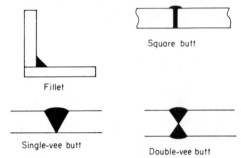

Fillet

Square butt

Single-vee butt

Double-vee butt

Fig. 4.150

Defects in welded joints

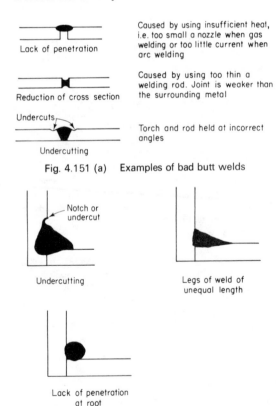

Lack of penetration

Caused by using insufficient heat, i.e. too small a nozzle when gas welding or too little current when arc welding

Reduction of cross section

Caused by using too thin a welding rod. Joint is weaker than the surrounding metal

Undercuts

Torch and rod held at incorrect angles

Undercutting

Fig. 4.151 (a)   Examples of bad butt welds

Notch or undercut

Undercutting

Legs of weld of unequal length

Lack of penetration at root

Fig. 4.151 (b)   Defects in fillet welds

Edge preparation for oxy-acetylene welding. Fig. 4.152 shows the way in which the edges of plates should be prepared for butt welding.

Blowpipe nozzles. When welding thick plate more heat is needed than when welding thin plate. This is made possible by increasing the size of the blowpipe nozzle. The best size of nozzle must always be used for any particular plate thickness. When too small a nozzle is used a high gas pressure is needed to produce the heat required. This causes the flame to leave the end of the nozzle making it practically impossible to obtain a good weld. If too large a nozzle is used the gas pressure must be reduced otherwise too much heat will be generated. This results in explosions at the nozzle making good welding impossible.

Welding rods. Welding rods are used in order to supply filler metal, thus ensuring that the joint is as strong as the metals being welded. Generally speaking the rod is made from the same metal as the plates being joined. Steel rods, for instance, contain 0·2% to 0·4% of carbon, 14% of manganese and 0·08% of silicon.

Fluxes. When welding steel no flux is needed because iron oxide has a lower melting temperature than steel. It floats to the surface of the molten metal and forms a slag which is easily removed. For most other metals borax and water form a suitable flux.

Flame cutting. Steel may be easily cut by using an oxy-hydrogen, oxy-propane, oxy-coal gas or an oxy-acetylene cutting blowpipe (Fig. 4.153). Oxygen and the other gas being used are mixed and supplied to nozzles A and B. When the torch is lit these gases burn and provide a flame which heats the metal to a bright-red heat. A stream of oxygen from nozzle C is directed on to the metal which immediately oxidizes. The melting temperature of the oxide is lower than that of the steel and hence the oxide melts and is blown away by the stream of oxygen.

The size of the torch varies with the thickness of the metal to be cut. Propane and coal gas give a lower flame temperature than does acetylene and it therefore takes longer to raise to the cutting temperature. Oxy-hydrogen is used where the ventilation is poor because the products of combustion are not so harmful as those of other gases.

Electric arc welding. When electricity passes through an air gap from one conductor to another intense heat is produced. The temperature of an arc jumping between two conductors is in the region of 3500 °C. In electric arc welding it is this arc which provides the source of heat.

During welding an electric current of high amperage and low voltage is fed into a metal electrode and the arc is drawn between the electrode and the metal to be joined (Fig. 4.154). The high temperature of the arc melts the parent metal and the end of the electrode (or welding rod). Small drops of molten metal from the rod are deposited and this unites with the metal being welded.

Coated electrodes. When a base wire is used as an electrode the arc is difficult to control and also the weld tends to be brittle and porous. When the electrode is coated the arc becomes more stable. The coating forms

Plate thickness	Edge preparation	Remarks

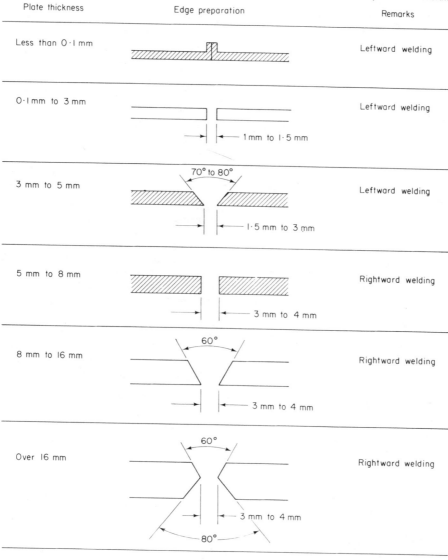

Fig. 4.152

a slag over the weld which protects the weld from the atmosphere and reduces the rate at which the weld cools, making it less brittle.

Preparation of plates for arc welding. Fig. 4.155 shows the way in which the edges of the plate should be prepared for butt welding.

Tack welding. When the plates are set up for welding they must be in perfect alignment. They are tack welded to hold them in this position. These tack welds are very short runs at intervals along the joint and should be made with a high current so that proper penetration is obtained.

The maximum pitch and length of the tack welds are obtained from the formulae:

Pitch = 100 mm + 16 × plate thickness in mm
Length of each tack = 3 × plate thickness

Hence, for a plate 12 mm thick:

Pitch = 100 + 16 × 12 = 292 mm (say 300 mm)
Length of each tack = 3 × 12 = <u>36 mm</u>

A.C. and D.C. applications. Both alternating (a.c.) and direct (d.c.) current are used for metal arc welding. Each has its own particular applications and in some cases either may be used. The advantages and disadvantages of both are given below:

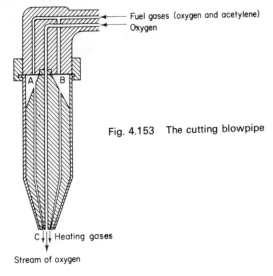

Fig. 4.153   The cutting blowpipe

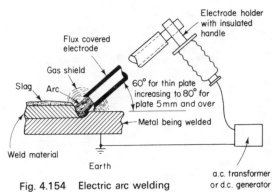

Fig. 4.154   Electric arc welding

(2) Little maintenance is required on a.c. equipment and it has a high electrical efficiency.

(3) D.C. equipment may be used with both ferrous and non-ferrous electrodes.

(4) D.C. equipment is especially suitable for welding sheet metal and pipes.

(5) In confined spaces and in damp conditions d.c. equipment is safer than a.c.

(1) The cost of a.c. equipment is less than the cost of d.c. equipment.

Thickness of plate	Plate preparation	Remarks
Up to 3·15 mm	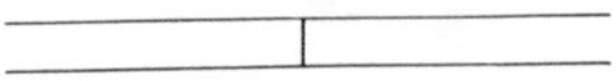	No preparation. Plates are butted together
3·15 mm to 6·30 mm	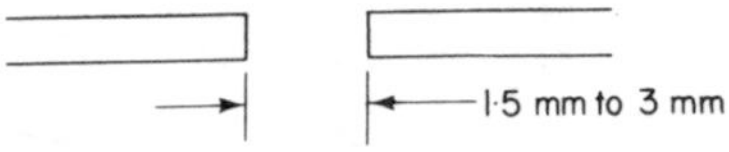 1·5 mm to 3 mm	No preparation of plates. Gap is needed for penetration
6·30 mm to 20·0 mm	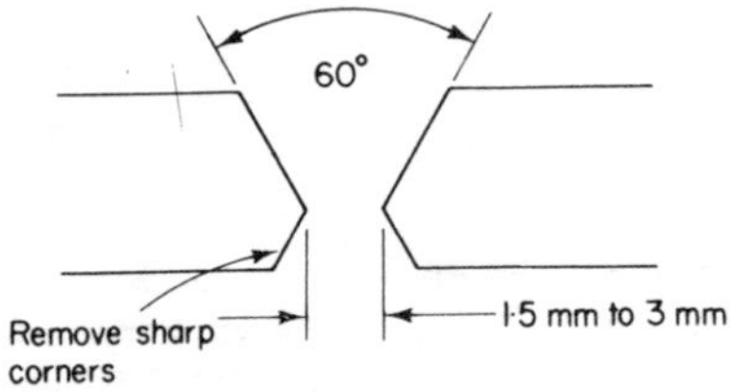 60° Remove sharp corners 1·5 mm to 3 mm	
Over 20·0 mm	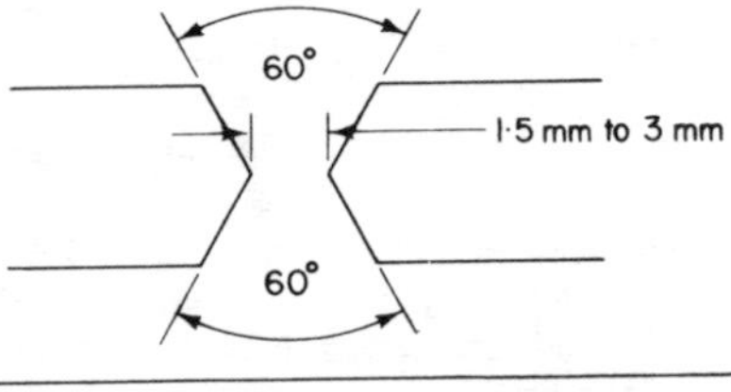 60° 1·5 mm to 3 mm 60°	

Fig. 4.155   Preparation of plates for arc welding

Safety

(1) Handle gas cylinders in the correct way as stated in the booklet 'Safety in the use of compressed gas cylinders' which is issued by the British Oxygen Company. Gas cylinders are painted different colours according to the gas contained in them as follows: oxygen—black; acetylene—maroon; coal gas—red; hydrogen—red; air cylinders—grey; nitrogen—grey with black neck.

(2) Make sure that all arc-welding equipment is properly maintained by skilled maintenance workers.

(3) Never chip slag whilst it is hot without wearing chipping goggles.

(4) When oxy-acetylene welding always wear goggles.

(5) When arc-welding always use the special welding visor provided. Failure to do so can result in severe eye injuries.

(6) Screens must be placed around the work when arc welding to protect passers-by.

(7) When a flux is used always use the extractor fan to give adequate ventilation.

Exercise 4.12

1. Give reasons why welding is often preferred to casting or riveting.
2. Three sorts of welding flame may be obtained when using oxy-acetylene. What are they? Describe how each type of flame is obtained and when it would be used.
3. What is the main difference between soldering and fusion welding?
4. Describe the process of leftward welding. When is this kind of welding used?
5. What is rightward welding? When should it be used in preference to leftward welding?
6. Sketch (a) a single-vee butt joint (b) a square-butt joint (c) a fillet joint.
7. Why is it important that the correct size of nozzle is used when oxy-acetylene welding?
8. Sketch three types of defect that can occur when fillet welding.
9. Show three defects in butt welds giving one reason why each occurs.
10. Briefly describe the process of flame cutting.
11. Why are coated electrodes used during electric arc welding?
12. Briefly describe how the heat source is obtained for electric arc welding.
13. Explain why a flux is not needed when welding steel by the oxy-acetylene process.
14. State three advantages that d.c. equipment has over a.c. equipment for electric arc welding.
15. Why is tack welding used?

4.13 Adhesives

We have seen in previous chapters that parts may be joined together by riveting, bolting, soldering, brazing and welding. Metal and plastic parts may also be joined by using adhesives.

Types of adhesives. Adhesion between surfaces can be either mechanical or specific. The adhesion is *mechanical* when there is an interlocking bond between the adhesive and the surfaces. In order to obtain a mechanical adhesion the surfaces must be porous enough to accept the glue. Traditional glues used for joining wood etc., worked on this principle.

The adhesion is *specific* when there is a molecular attraction between the adhesive and the surface. Adhesives used for joining metals and plastics depend mainly upon specific adhesion although some mechanical adhesion will also be present. In effect, with specific adhesion, the adhesive becomes an integral part of the surface. Metals and plastics usually have smooth surfaces and, in general, the smoother and less porous the surface the more the bond depends upon specific adhesion.

Thermosetting resins (e.g. epoxy resins) are used extensively for metal-to-metal bonding. Rubber-based adhesives are used for bonding friction linings and other hard, non-porous surfaces, to metal. There are a great number of different types of adhesives which are used for various purposes (see for instance Mechanical World Year Book 1970).

Choice of adhesives. The choice of adhesive depends upon many factors the most important of which are:
(1) The adhesive must be suitable for both of the materials to be joined together.
(2) For structural work the size and type of stress must be considered.
(3) At high temperatures all adhesives lose some of their strength and hence the temperature the adhesive will meet in service must be considered.
(4) Some substances such as oil, steam, petrol and chemicals attack the adhesive. Therefore, the substances with which the adhesive will come into contact during service must also be considered.
(5) The temperature and pressure required for the application of the adhesive and the form of the adhesive are also important considerations.

Strength of adhesive joints. For maximum strength, joints must be made so that the load which the joint has to carry is evenly distributed over the bonded area. This condition is only obtained if the joint is in tension or shear (Fig. 4.156). If cleavage or peel is present the joint is badly designed and weak (Fig. 4.157).

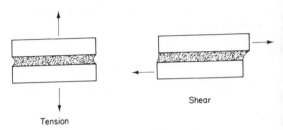

Fig. 4.156   Well designed joints must be in tension or shear

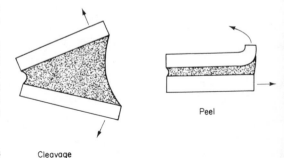

Fig. 4.157   Examples of badly designed joints

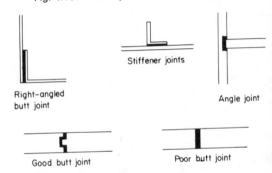

Fig. 4.158   Types of joints that can be made by using adhesives

Surface preparation of materials. Joints having good strength are usually obtained after the removal of grease and loose surface contamination. A light abrasive treatment of the surface helps, as a roughened surface gives a better key than do smooth surfaces. A wire brush, emery cloth or grit blasting is suitable.

Types of joints. Some of the joints that can be made by using adhesives are shown in Fig. 4.158.

Exercise 4.13
1. State three important factors which affect the choice of an adhesive.
2. Why does a roughened surface help when using adhesives? How is the surface prepared?
3. What is peel?
4. State the stress conditions for a good joint.
5. Sketch three joints that can be made by using adhesives.
6. What is mechanical adhesion?
7. What is specific adhesion? For metal-to-metal bonding why must the adhesion be specific?

4.14   Pipework

Pipes are used for carrying liquids (e.g. oil and water) and gases (e.g. steam and compressed air). Various materials are used, e.g. steel, copper, plastic and rubber. The choice of the pipe material depends upon the nature of the fluid which is to be used in the pipe, the pressure at which the pipe is to be used and the amount of flexibility required.

Types of pipes
1. *Steel* pipes are used for carrying cold, non-corrosive liquids and for steam. Three kinds of tube are available, light, medium and heavy which can be identified by a colour band as follows:
    Light tubes—brown colour band
    Medium tubes—blue colour band
    Heavy tubes—red colour band
The table opposite will give some idea of the sizes that are available.
2. *Copper* pipes are used for carrying water, steam, petrol, compressed air, etc. Pressures up to about 700 kN/m² and temperatures up to 100 °C may be accommodated by copper pipes. The pipes are usually seamless and they have the advantage that they are easily bent and joined together. Copper pipes do not corrode either from the fluid being carried inside the pipes or from the atmosphere surrounding them.
3. *Plastic* pipes are frequently used for low-pressure work. A typical example is the pipe which carries

petrol between the tank and the carburettor of a motor mower. Nylon and polythene are commonly used.

SIZES OF STEEL PIPES

Bore (mm)	Light Thickness (mm)	Medium Thickness (mm)	Heavy Thickness (mm)
6	1·8	2·0	2·65
8	1·8	2·35	2·9
10	1·8	2·35	2·9
15	2·0	2·65	3·25
20	2·35	2·65	3·25
25	2·65	3·25	4·05
32	2·65	3·25	4·05
40	2·9	3·25	4·05
50	2·9	3·65	4·5
65	3·25	3·65	4·5
80	3·25	4·05	4·85
90	3·65	4·05	4·85
100	3·65	4·5	5·4
125	—	4·85	5·4
150	—	4·85	5·4

Joining pipes together. There are many ways of joining pipes together, some of which are as follows:

1. Flanged joints. High-pressure flanged joints are made as shown in Fig. 4.159.

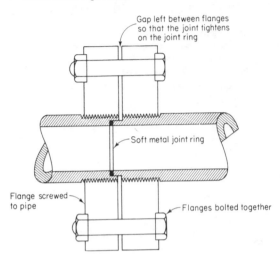

Fig. 4.159   High-pressure flanged joint. (A pressure of about 17 000 kN/m² may be carried)

Low-pressure flanged joints are made like Fig. 4.160. The flanges may be welded or screwed to the pipe ends. When the flanges are screwed on, a jointing-compound (often in the form of a paste) is used to seal the thread. The joint relies on contact between the two faces of the flanges to prevent leakage. However, the faces of the flanges may be irregular and jointing material is used to seal the joint. Soft jointing is used for water services at pressures up to 1700 kN/m² and is usually made from synthetic or pure rubber. Hard jointing made from asbestos fibre bonded with rubber is used for high-pressure and high-temperature work. Metal jointing of copper, brass or monel metal is used for steam pipelines employed in power installations.

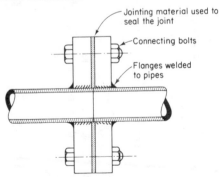

Fig. 4.160   Flanged coupling for low and medium pressures

2. Non-manipulative compression joints (Fig. 4.161). This joint does not require any working of the pipe except to cut it square. The joint is made tight by means of the soft copper ring which is compressed between the coupling and the pipe.

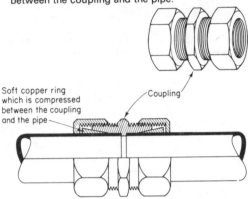

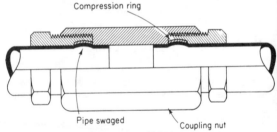

Fig. 4.161   Non-manipulative compression joint

3. Manipulative compression joint (Figs 4.162 and 4.163). The tube is swaged or flared by using special tools.

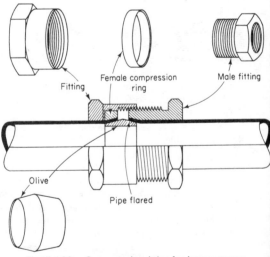

Fig. 4.162.   Compression joint for compressed air lines

Fig. 4.163   Compression joint for heavy-gauge copper pipe

The coupling nuts cause pressure to be exerted on the swaging or flaring thus making the tube mate with the compression ring. In copper tube pressures up to about 1000 kN/m² may be accommodated. Steel compression joints can withstand pressures up to about 40 000 kN/m².

4. *Capillary soldered joints* (Fig. 4.164). This is essentially a sweated joint which is frequently used in industrial and domestic heating systems. The closeness of fit between the tubes and the fitting is of the utmost importance. It is this which makes sure that the molten solder creeps over the joint area. Frequently the fittings are ready soldered by the manufacturer.

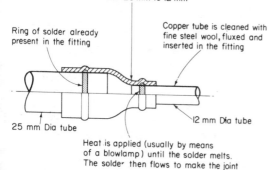

Fitting for reducing tube Dia from 25 mm to 12 mm

Ring of solder already present in the fitting

Copper tube is cleaned with fine steel wool, fluxed and inserted in the fitting

25 mm Dia tube

12 mm Dia tube

Heat is applied (usually by means of a blowlamp) until the solder melts. The solder then flows to make the joint

Fig. 4.164   Reading soldered joint for light-gauge copper tube

5. *Brazed joints.* These are often used for joining copper pipes, especially those of larger diameter. Brass flanges are often brazed to copper pipes when a flanged joint is needed.

6. *Special fittings* (Fig. 4.165) are obtainable for many purposes.

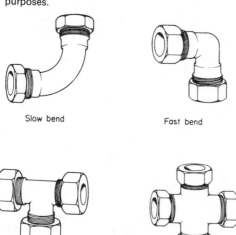

Slow bend

Fast bend

Tee piece

Four way

Fig. 4.165   Special pipe fittings

Pipe threads. The form of the standard pipe thread is the ISO metric form. The size of the thread is given according to the *bore* of the pipe on which the thread is to be cut. The thread may be parallel or tapered. The tapered threads are used with fittings which have parallel threads. Tapered threads are self sealing when pulled up tight.

Bending of pipes. Copper tubes can easily be bent. The original length of the pipe remains unaltered after bending only along the centre-line of the pipe. The inside of the bend is shortened whilst the outside is lengthened. This shortening and lengthening is likely to produce a shape similar to Fig. 4.166 unless steps are taken to prevent it occurring. If this shape does occur after bending the pipe is likely to collapse in service.

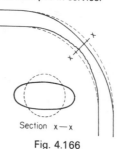

Section x — x

Fig. 4.166

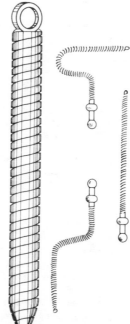

Section of spring

Fig. 4.167   Polished steel springs for bending light gauge copper tubes

When bending pipes manually the tube walls must be supported whilst bending takes place. This support may be provided by using

(1) lead;
(2) pitch or resin;
(3) sand;
(4) steel springs (Fig. 4.167).

Machine bending. When a machine is used for bending (Fig. 4.168) special formers or mandrels are used to support the walls of the pipes. Fig. 4.169 shows a small hand bending machine suitable for tubes up to 12 mm diameter.

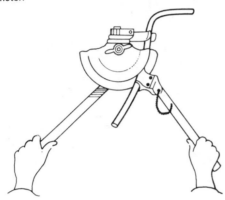

Fig. 4.168   Small hand bending machine

Fig. 4.169 shows a draw-bar bending machine with a stationary mandrel which is used for making small radius bends.

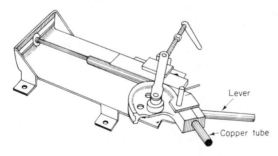

Fig. 4.169   Draw-bar bending machine

There are many other types of bending machines available which will allow copper pipes up to 63 mm diameter to be bent without loading with sand etc.

Heat treatment of pipes before bending. Copper pipes should be annealed before attempting to bend them. The pipe need only be annealed over the length of the bend. The pipe is heated to a dull-red heat (600 °C). It can be cooled by plunging it into cold water or it can be

left to cool naturally. The method of cooling has no effect on the degree of softness obtained.

Steel pipes are usually forge-bent whilst red hot although they can also be bent by machine. These pipes are usually loaded with sand prior to the heating and bending.

Colour coding. All pipework should be colour coded after it has been installed. The colour code indicates the medium flowing in the pipe which allows for safe and convenient maintenance. The recommended colours are shown below and have been taken from BS 1710.

COLOUR CODE FOR PIPEWORK

Pipe contents	Colour
Compressed air under 1350 kN/m²	White
Compressed air over 1350 kN/m²	White with a red band
Steam	Aluminium or crimson
Town gas	Canary yellow
Diesel fuel	Light brown
Lubricating oil	Salmon pink
Hydraulic-power oil	Salmon pink
Chemicals	Dark grey
Boiler-feed water	Sea green
Drinking water	Aircraft blue
Central-heating water	French blue
Water for hydraulic power	Mid-Brunswick green

For a full list consult the specification BS 1710.

Exercise 4.14

1. State four materials from which pipes are made.
2. State three factors which affect the choice of material from which a pipe is made.
3. On steel pipes a colour band is used to identify the type of pipe. What colour band is used for: (a) light tubes; (b) medium tubes; (c) heavy tubes?
4. State four liquids or gases which may be carried in copper pipes.
5. What is the purpose of a jointing material when it is used for a flanged joint?
6. Name three kinds of jointing material and state what each is used for.
7. By means of sketches show the difference between a non-manipulative joint and a manipulative joint.
8. What is a capillary-soldered joint?
9. How are steel pipes protected against corrosion?
10. Why are steel pipes loaded? What materials are available for loading?
11. Why are copper pipes annealed before bending?
12. Why are pipes which are to be machine-bent left unloaded?

4.15   The lathe

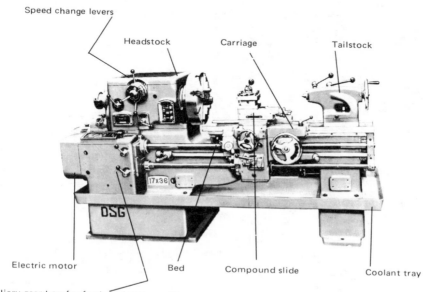

Speed change levers

Headstock   Carriage   Tailstock

Electric motor   Bed   Compound slide   Coolant tray

Auxiliary gear box for feeds and screw cutting

Fig. 4.170   *(Dean Smith & Grace Ltd.)*

The lathe is the most versatile of machine tools. It can be used for a number of cutting operations, as described later in this chapter. Generally, the lathe holds the work and turns it under power. The tool is fed into the moving work to remove metal. Cylindrical surfaces, both internal and external, flat surfaces, screw threads and slots can all be cut on the lathe. The main parts of a lathe are shown in Fig. 4.170.

Main parts of the lathe

The lathe bed. Fig. 4.171 illustrates the main details.

The machined guideways may be flat, vee or a combination of flat and vee. The flat and vee type is illustrated in Fig. 4.171 and the flat type is shown in Fig. 4.172.

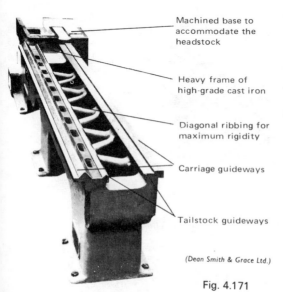

Machined base to accommodate the headstock

Heavy frame of high-grade cast iron

Diagonal ribbing for maximum rigidity

Carriage guideways

Tailstock guideways

(Dean Smith & Grace Ltd.)

Fig. 4.171

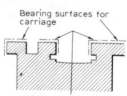

Bearing surfaces for carriage

Fig. 4.172

Tailstock bearing surfaces

Handle for clamping
the tailstock to the
lathe bed

Rotating the handwheel
causes the barrel to move

Handle for clamping
the barrel

Barrel which has an
internal Morse taper
to accommodate a
centre or tools

Machined guideways

Two screws, one on
either side of the
tailstock, which
allow the tailstock
to be moved a short
distance across the
baseplate

(Dean Smith & Grace Ltd.)

Fig. 4.173   Tailstock

Notice that for both types the tailstock slides on one pair of surfaces whilst the carriage slides on the other pair. This arrangement prevents wear on any one part of the bed.

The headstock contains all the gears and mechanisms necessary to obtain a suitable range of spindle speeds. Most machines now use a geared type of headstock.

The tailstock (Fig. 4.173) is used to support one end of the work when it is being turned between centres. It is also used to carry such tools as drills and reamers.

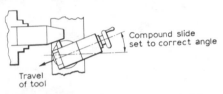

Compound slide
set to correct angle

Travel
of tool

Fig. 4.174 (a)   Setting of compound slide

The saddle or carriage, Fig. 4.174, is a casting machined so that it slides on the bedways.

The calibrated dial is used when accurate work is needed. It allows the tool to be fed into the work with high precision. The operator must know the value of each division on the dial. The following examples show how the calibrated dial can be checked and used.

The cross-slide moves at right angles to
the bedways. It is operated by a left-hand
screw and nut and handwheel.
The cross-slide is moved by the handwheel
or by an automatic drive

The compound top slide may be swivelled
to give any angle which is desired.
This allows the tool to be moved in
directions other than those permitted
by the carriage and the cross-slide
movements (see diagram (a))

Leadscrew, used when screw
threads are to be cut

Feed shaft used when automatic
feed is required

The calibrated dial is used when it is
required to move the cross-slide through
an exact amount.
This is often needed when setting a tool
to give an accurate depth of cut

The saddle is moved along the bed by rotating
this handwheel. It can also be moved by
using an automatic drive

The apron which contains mechanisms for
moving the saddle and the cross-slide

(Dean Smith & Grace Ltd.)

Fig. 4.174 (b)   Details of lathe carriage

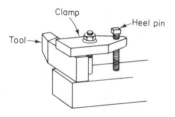

Clamp type. The height of the tool is adjusted by using packing pieces under the tool which is inconvenient

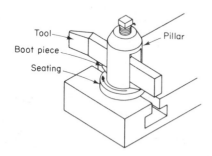

American pillar type. The tool height is adjusted by rocking the boot piece. The effective cutting angles on the tool are altered by the adjustment to the boot piece

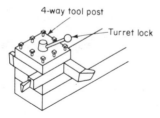

Four-way tool post allows four tools to be placed in the tool post thereby allowing a sequence of operations to be carried out without changing the tools. The post is swivelled through 90° when a change of tools is needed

Fig. 4.175   Tool posts

Example. When checking the calibrated dial of the cross-slide it is found that five revolutions of the handwheel makes the cross-slide move 25 mm. The dial has 100 divisions. Find:

(a) The distance moved by the cross-slide for each division of the dial;

(b) The number of divisions needed to give the cross-slide a movement of 0·50 mm.

(a) Pitch of the screw = $\dfrac{25}{5}$ = 5 mm

Number of divisions on the dial = 100
Distance moved by the cross-slide for each division on the dial = $\dfrac{5}{100}$ = 0·05 mm

(b) Movement required = 0·50 mm

Divisions needed   = $\dfrac{0·50}{0·05}$ = 10 divisions

Example. It is known that each division of the calibrated dial of a cross-slide gives a movement of 0·02 mm. If the reading on the dial is 27 and an inward cross-slide movement of 0·28 mm is needed, what will be the reading on the dial when this movement is achieved?

Divisions on dial needed = $\dfrac{0·28}{0·02}$ = 14

Dial reading has to be 14+27 = 41 divisions

Tool posts. The three types of tool post most commonly used are shown in Fig. 4.175. Note the way in which the posts restrict the movement of the tool.

Lubrication of the lathe. The guideways must be kept lubricated. An oil pump in the headstock circulates oil to the gears and bearings. An oil-level gauge ensures that the correct oil level is maintained. All the guideways should be oiled daily. The apron should be topped up with oil as indicated on the oil-level gauge.

The cross-slide is oiled through a nipple on the top of the slide.

Action of the lathe, Fig. 4.176. When the lathe is in operation the work is held firmly either between centres or in a chuck and made to revolve. The tool is locked in the tool post and fed into the work (see Fig. 4.176 (i) and (ii)) by hand or power according to the type of cutting operation.

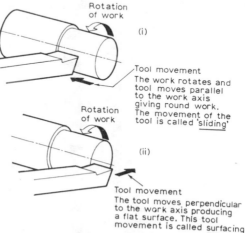

Fig. 4.176   The action of a lathe

Turning between centres. The work is first prepared by drilling each end with a centre drill. This is fully described on page 187.

Two centres are needed to support the work. The headstock centre revolves with the lathe spindle. It must

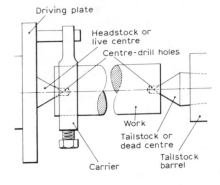

Driving plate

Headstock or live centre

Centre-drill holes

Work

Tailstock or dead centre

Tailstock barrel

Carrier

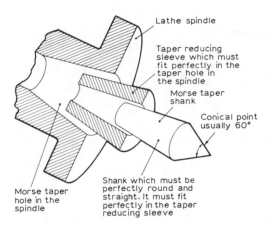

Lathe spindle

Taper reducing sleeve which must fit perfectly in the taper hole in the spindle

Morse taper shank

Conical point usually 60°

Shank which must be perfectly round and straight. It must fit perfectly in the taper reducing sleeve

Morse taper hole in the spindle

Fig. 4.179   Assembly of the live centre

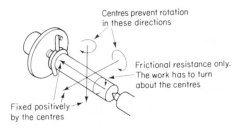

Centres prevent rotation in these directions

Frictional resistance only. The work has to turn about the centres

Fixed positively by the centres

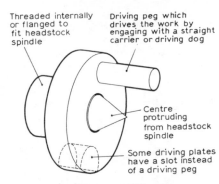

Threaded internally or flanged to fit headstock spindle

Driving peg which drives the work by engaging with a straight carrier or driving dog

Centre protruding from headstock spindle

Some driving plates have a slot instead of a driving peg

Fig. 4.180   The driving plate

Centre made from carbon steel hardened and ground

Centre angle 60° (may be 90° for heavier work)

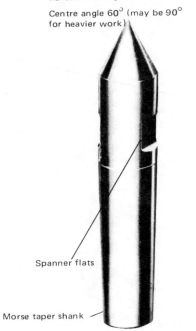

Spanner flats

Morse taper shank

(A.A. Jones & Shipman Ltd.)

Fig. 4.178   A lathe centre

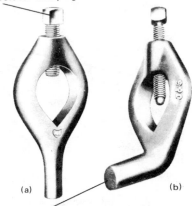

Bolt for clamping on to the work

(a)

(b)

Tail fits into slot in the driving plate. Care must be taken to make sure that there is clearance between the bottom of the slot and the carrier

Fig. 4.181   (a) Straight carrier, used where the driving plate has a peg.
(b) Bent tail carrier, used where the driving plate has a slot.

run true, otherwise it is impossible to turn work so that it is parallel from end to end. The tailstock centre is stationary, hence the name *dead centre*. Fig. 4.177 shows how the work is supported. Note how the work is prevented from moving whilst the cut is being taken.

Fig. 4.178 shows a lathe centre and Fig. 4.179 shows how the live centre is fitted.

The *driving plate*, Fig. 4.180, is used to drive the work which is supported on centres.

Carriers or driving dogs are shown in Fig. 4.181.

Chucks. Two types are used—the three-jaw self-centring type and the four-jaw type.

The *three-jaw chuck*, Fig. 4.182, is used for gripping round and hexagonal work. Hard jaws are used where it does not matter if the work is marked. Soft jaws are used to prevent the work from being damaged and they are bored out to suit the work. When the jaws are worn the work will run slightly out of true. No adjustment is possible but only with very accurate work is this important.

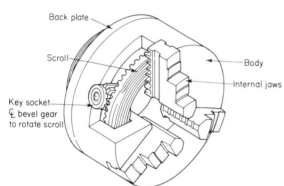

Fig. 4.182   Three-jaw chuck

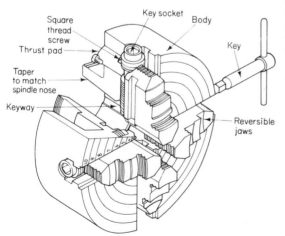

Fig. 4.183   Four-jaw chuck. Each jaw is operated separately by its own square-threaded screw

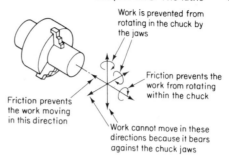

Fig. 4.184   The restraints provided by a chuck

In the *four-jaw chuck*, Fig. 4.183, each of the jaws operates independently. It gives a better grip than the three-jaw type and it can accommodate various shapes of work.

Fig. 4.184 shows how the work is restrained from moving by the chuck.

With both types of chuck, work should never be removed until it is certain that all the operations have been completed. Once work has been removed from a chuck it is extremely difficult to reset it so that it runs true.

Fitting and removal of chucks

(1) Make sure that the machine is switched off at the mains.

(2) Place a piece of wood on the guideways to prevent them from being damaged if the chuck should fall.

(3) Make sure that the locating faces of the spindle and the chuck are clean and free from swarf before fitting a chuck.

Lathe turning tools may be made from plain carbon steel or high-speed steel. Quite often the tool consists of a high-speed-steel tip welded to a toughened steel shank. Tungsten carbide and ceramic tips are used in the same way. The size of a lathe tool is determined by the depth of shank, width of shank and overall length.

Fig. 4.185 shows the names given to the various parts of a lathe tool.

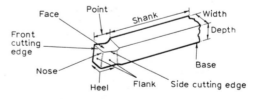

Fig. 4.185

Figs 4.186 (a) to (f) show some commonly-used turning tools and their applications.

Lathe tool cutting angles. All turning tools have front and side clearance angles. These prevent the tool from

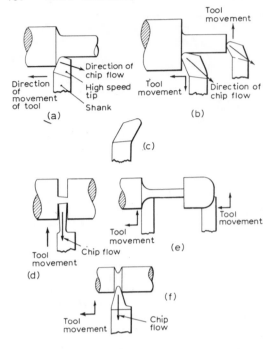

Fig. 4.186   Turning tools:
(a) Right-hand straight roughing tool; (b) Right-hand turning and facing tool; (c) Left-hand tool; (d) Parting-off and recessing tool; (e) Form tools; (f) Screw-cutting and vee tool

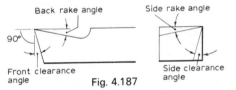

Fig. 4.187

rubbing on the work, clearance angles of 5°–10° are satisfactory. The tool can have either back or side rake or both. The correct rake angle allows the chips to slide easily over the tool face. Too large a rake angle weakens the tool and causes it to pull into the work. Soft and ductile metals require larger rake angles than do brittle metals.

The recommended rake angles for lathe tools are shown in the table below.

Metal being cut	Back rake	Side rake
Soft steel	8°	20°
Mild steel	8°	18°
High-carbon steel	8°	14°
Cast iron	8°	14°

Plan approach angle. This is the angle made by the cutting edge of the tool to the work; see Fig. 4.188. Too large an approach angle may cause chatter and vibration.

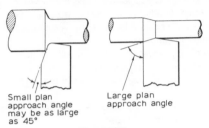

Fig. 4.188

Setting the tool in the lathe. Fig. 4.189 illustrates the importance of setting the tool at the correct height.

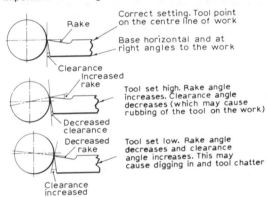

Fig. 4.189   Effect of tool height on rake and clearance angles

Cutting speeds. The cutting speed for turning varies according to the material being cut and the material from which the tool is made. Some idea of the cutting speeds used, when cutting with high-speed steel tools, can be obtained from the following table.

CUTTING SPEEDS FOR HIGH-SPEED STEEL TOOLS

Metal being cut	Cutting speed metres/min
Cast iron	15
Mild steel	25
Tool steel	15
Brass	90
Aluminium	180

Cutting fluids. These are used in machining operations because:
(1) They prevent the tool and work from becoming overheated by carrying away much of the heat generated by the cutting operation. Higher cutting speeds may also be used.

(2) They lubricate between the chip and the top face of the tool. This saves power and prolongs the life of the tool and also gives a better finish to the work.

(3) They wash away the chips, thus preventing the tool point from becoming clogged.

Cutting fluids may consist of fixed oil, mineral oil, or a mixture of oil and water. The *fixed oils* consist of animal, fish and vegetable oils such as lard oil, whale oil, cottonseed oil and linseed oil. The *mineral oils* are derived from crude petroleum, paraffin being an example. The vegetable oils lubricate better than the mineral oils but they tend to become gummy when heated. *Compounded oils* consist of a mixture of mineral and vegetable oils and sometimes sulphur is added to form *sulphonated oil* which forms a highly adhesive film on the metal. The most common cutting fluids are the *soluble oils* which are a mixture of oil and water and are known as 'suds'. These have better cooling properties but they do not lubricate as well as the oils.

The fluids recommended are as follows:

For cutting mild steel—soluble oil in proportions varying from 1 to 10 to 1 to 100.

For cutting tool steel—sulphonated or compounded oil.

For cutting aluminium—paraffin.

For cutting brass and bronze—soluble oil.

Cast iron is usually cut dry.

Roughing and finishing cuts. At least two cuts should be taken—a roughing cut and a finishing cut. Unless a finishing cut is taken the work dimensions may not be accurate as the work expands during the roughing cut due to heat generation.

The depth of the roughing cut depends upon the amount of metal to be removed, the size and strength of the machine and the rigidity of the work. When a large amount of material is to be removed, the depth of cut should be as large as the machine will take, with a coarse feed. The work should be turned for about 12 mm and checked with calipers and rule before completing the cut. About 1 to 2 mm should be left on the diameter for the finishing cut and when turning up to a shoulder the cut should be stopped when within 1 to 2 mm of the required length.

After the roughing cut a finishing tool is set up. More care is needed and more attention must be paid to measurement of length and diameters. Important diameters may be checked by micrometer or caliper gauge and lengths by vernier calipers. A fine feed is used to provide a good finish.

The work should be finished with the tool. The use of emery cloth is permissible when a good surface finish is needed. Files should never be used.

The centre holes needed for supporting work mounted on centres. Before the work can be mounted on the centres the ends of the work must be drilled. A

combination centre drill is used for this purpose, Fig. 4.190 (a).

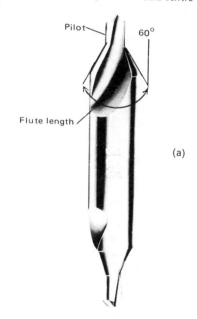

Angle of centre hole must be the same as the angle on the lathe centre

Pilot   60°

Flute length

(a)

Parallel part is held in a drill chuck

(A.A. Jones & Shipman Ltd.)

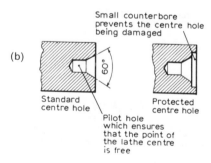

(b)

Small counterbore prevents the centre hole being damaged

60°

Standard centre hole

Protected centre hole

Pilot hole which ensures that the point of the lathe centre is free

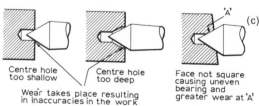

(c)

'A'

Centre hole too shallow

Centre hole too deep

Wear takes place resulting in inaccuracies in the work

Face not square causing uneven bearing and greater wear at 'A'

Fig. 4.190

It is very important that the centre holes are made correctly. Some of the faults arising from faulty centre holes are shown in Fig. 4.190 (c).

Methods of centring. Three methods are commonly used: (1) Centre finder method, Fig. 4.191. The vee of the centre finder is held against the end face and a line is scribed on the work. The work is then rotated about 90° and a second line is scribed. The intersection of these lines is the centre of the work.

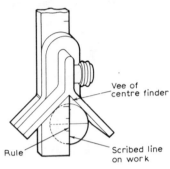

Fig. 4.191

(2) Bell centre punch method, Fig. 4.192. In this method the cone of the bell is placed over the end of the work and the centre punch is then given a light tap. The cone automatically positions the centre punch.

(3) Marking out method.

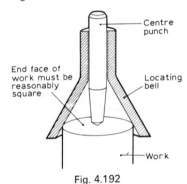

Fig. 4.192

Drilling the centre holes. The commonest method is to drill the holes on the lathe as shown in Fig. 4.193. The

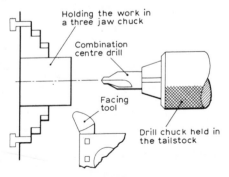

Fig. 4.193

work is held in a 3-jaw chuck and the end is faced before drilling. The work is then reversed in the chuck and the other end is faced and drilled.

Centre holes can also be drilled in a drilling machine by holding the work vertically in the machine vice.

Checking the lathe centres. If the lathe centres do not line up properly it is impossible to turn work parallel. The method of checking the centres and the effects of centres which do not run true or do not line up are shown in Figs 4.194 and 4.195.

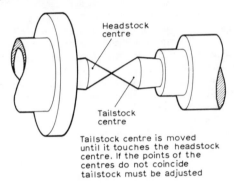

Fig. 4.194   Checking the alignment of the centres

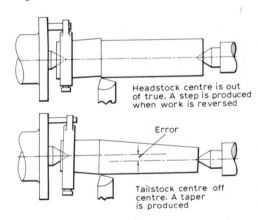

Fig. 4.195   Faults arising from misalignment of the lathe centres

Boring. A bored hole is usually more accurate than a hole which has been drilled and reamed. This is because the drill may run out a little and the hole may then not be true with the axis of the work. Accurate holes are frequently bored and finished by reaming. In this case the amount removed by the reamer will be only 0·05 mm to 0·1 mm and the reamer will follow the bored hole. This practice is to be recommended because it speeds up the process, as there is no fear of taking too deep a cut during the boring operation.

Boring may be done with a boring tool or a boring bar, both of which are illustrated in Fig. 4.196. The diameter of the tool shank or the boring bar should be as large as possible because greater rigidity is then obtained. Boring tool clearances are greater than with external cutting tools according to the size of the hole.

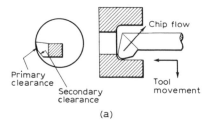

Primary clearance

Secondary clearance

Chip flow

Tool movement

(a)

Square section for clamping boring bar in the tool post

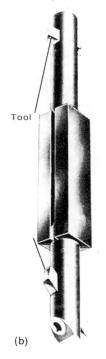

Tool

(b)

(A.A. Jones & Shipman Ltd.)

Fig. 4.196   (a) Boring tool   (b) Boring bar

Mandrels. It frequently happens that work is finish-bored before the outside is turned. In this case the work can be mounted on a *mandrel* so that the outside can be finish-turned. A mandrel is simply a shaft which has centre holes provided. The centre holes are usually lapped for greater accuracy and their ends are recessed to protect them from damage. It is most important that the diameter is perfectly true with the centre holes.

These mandrels depend upon frictional contact for driving work against the cut, therefore they are only used for light cuts with the feed of the tool traversing towards the large end.

Besides finish-turning operations mandrels are also used for grinding, milling and inspection to check concentricity.

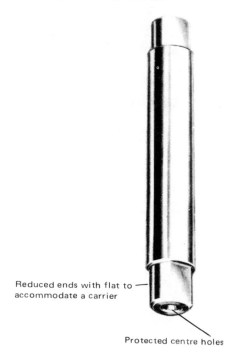

Mild steel case hardened and ground to a good finish

Reduced ends with flat to accommodate a carrier

Protected centre holes

(A.A. Jones & Shipman Ltd.)

Fig. 4.197   Plain mandrel

Expanding mandrel, Fig. 4.198:
Expanding mandrels are supplied in sets and they thus give greater economy by reducing the number of mandrels needed. They allow the work to overhang the sleeve thereby allowing the ends of the work to be faced. Because they give a better grip, heavier cuts may be taken.

Faceplates. When work has to be turned parallel to or at right-angles to a flat surface a faceplate is used (Fig. 4.199).

If the work is not symmetrical the out-of-balance forces will damage the spindle bearings and also produce chatter when cutting, resulting in a poor finish. A balance weight (Fig. 4.200) is used so that the centre of gravity of the faceplate and work is at the geometrical centre of the faceplate. Out-of-balance forces are then prevented.

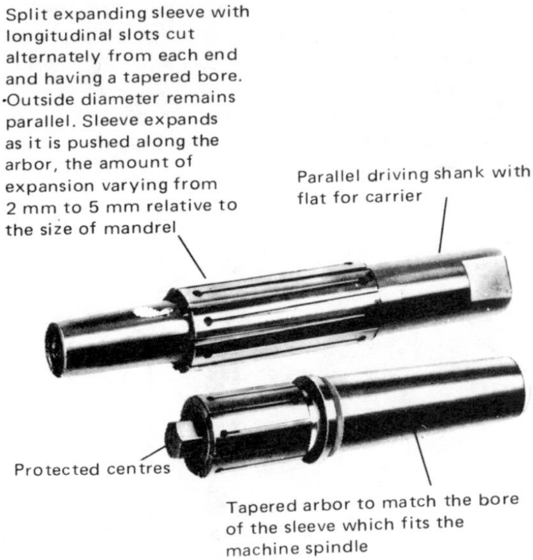

Split expanding sleeve with longitudinal slots cut alternately from each end and having a tapered bore. ·Outside diameter remains parallel. Sleeve expands as it is pushed along the arbor, the amount of expansion varying from 2 mm to 5 mm relative to the size of mandrel.

Parallel driving shank with flat for carrier

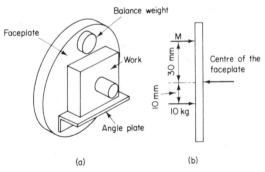

Protected centres

Tapered arbor to match the bore of the sleeve which fits the machine spindle

(Alfred Herbert Ltd.)

Fig. 4.198　Expanding mandrel

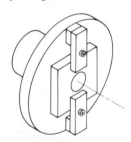

Fig. 4.199　Work clamped to a face plate

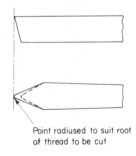

Balance weight

Faceplate

Work

M

30 mm

10 mm

10 kg

Centre of the faceplate

Angle plate

(a)　　　　　(b)

Fig. 4.200　A balance weight is used so that the centre of gravity of the faceplate and the work is at the geometrical centre of the faceplate

The following example shows how the mass of a balance weight may be calculated.

Example. In Fig. 4.200 (a) the c.g. of the work lies 10 mm below the centre of the boss and the work has a mass of 10 kg. A balance weight is to be fitted 30 mm above the ·centre of the faceplate. Find the mass of the balance weight.

Fig. 4.200 (b) shows the conditions. Taking moments about the centre of the faceplate:

$$M \times 30 = 10 \times 10$$
$$M = \frac{100}{30} = 3\tfrac{1}{3} \text{ kg}$$

The balance weight must have a mass of $3\tfrac{1}{3}$ kg

Screw cutting. Single-point tools are frequently used for cutting screw threads on the lathe. The most important features of a screw thread are:

(a) the form or shape of the thread;

(b) the pitch of the thread.

The shape of the cutting tool (Fig. 4.201) controls the form of the thread. The pitch of the thread is basically obtained by copying the pitch of the lathe leadscrew.

Most lathes have a leadscrew with the pitch accurately machined. When cutting a screw thread the leadscrew transmits a linear motion to the carriage by means of a split nut. When the leadscrew is given one revolution the carriage, and hence the tool, moves a distance equal to the pitch of the leadscrew.

When cutting a screw thread, the carriage must move a distance equal to the pitch of the thread to be cut whilst the work makes one revolution. Hence the speed of rotation of the leadscrew must be set relative to the speed of rotation of the spindle. On modern lathes this is done by using the gear box.

Point radiused to suit root of thread to be cut

Fig. 4.201　A screw-cutting tool

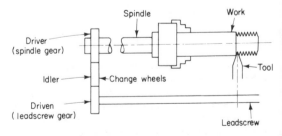

Spindle　　　Work

Driver
(spindle gear)

Idler　　Change wheels

Driven
(leadscrew gear)

Tool

Leadscrew

Fig. 4.202　The change-wheels for cutting a screw thread

On some of the older machines a gear train has to be set up between the spindle and the leadscrew as shown in Fig. 4.202.

The gear train may be calculated by using the formula:

$$\frac{number\ of\ teeth\ on\ spindle\ gear}{number\ of\ teeth\ on\ leadscrew\ gear}$$

$$= \frac{pitch\ of\ thread\ to\ be\ cut}{pitch\ of\ leadscrew}$$

The standard gears supplied with the machine have 20 teeth, 25 teeth, 30 teeth and so on in steps of 5 teeth up to 120 teeth. One of the smaller wheels is usually duplicated and this is often the 40-tooth wheel. On metric lathes the leadscrew usually has a pitch of 6 mm.

Example. Find suitable change-wheels for cutting a thread having a pitch of 1·5 mm. The lathe has a 6 mm pitch leadscrew.

$$\frac{\text{Number of teeth on spindle gear}}{\text{Number of teeth on leadscrew gear}} = \frac{1 \cdot 5}{6} = \frac{1}{4}$$

We must now convert this fraction by multiplying top and bottom by the same amount to suit the change-wheel sizes. Since the smallest gear has 20 teeth we may multiply top and bottom by 20.

$$\frac{\text{Number of teeth on spindle gear}}{\text{Number of teeth on leadscrew gear}} = \frac{1 \times 20}{4 \times 20} = \frac{20}{80}$$

Thus a suitable gear train is a <u>20-tooth wheel</u> driving an <u>80-tooth wheel</u>. The idler, which is used so that the spindle gear and the leadscrew gear rotate in the same direction, may have any convenient number of teeth. Note that this is not the only suitable gear train. By multiplying top and bottom of the fraction by 25 we obtain $\frac{1 \times 25}{4 \times 25} = \frac{25}{100}$ which is also satisfactory.

Example. Find suitable change-wheels for cutting a thread with a pitch of 1·25 mm on a lathe with a 6 mm leadscrew.

$$\frac{\text{Number of teeth in driver wheel}}{\text{Number of teeth in driven wheel}} = \frac{1 \cdot 25}{6}$$

To make the numerator (top of fraction) into a whole number multiply top and bottom by 4.

$$\frac{\text{Number of teeth in driver wheel}}{\text{Number of teeth in driven wheel}} = \frac{1 \cdot 25 \times 4}{6 \times 4} = \frac{5}{24}$$

$$= \frac{5 \times 5}{24 \times 5} = \frac{25}{120}$$

The change-wheels all have numbers of teeth which are divisible by 5 and hence we always multiply the top and the bottom of the fraction by either 5, 10, 15, etc., in order to obtain suitable gears.

The most popular way of cutting a screw thread is by using the angular feed method shown in Fig. 4.203. The tool cuts only on one side and any tendency for the thread to tear is thus prevented.

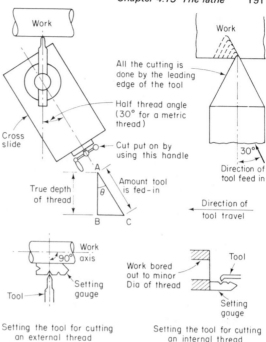

Fig. 4.203 Angular feed method for cutting a screw thread

The tool must be set absolutely square to the work. A setting gauge is used as shown in Fig. 4.203. The tool must also be set at the centre height of the work otherwise the wrong thread form will be produced.

When the crest of the thread is to be rounded a chaser is used (Fig. 4.204), since it is not possible for a single-point tool to round the crest.

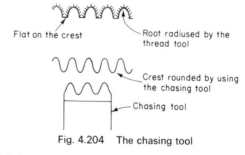

Fig. 4.204 The chasing tool

Safety

(1) Never use a lathe until you have been instructed how to use it.

(2) Be sure you know how to stop the machine quickly.

(3) Do not attempt to clean the machine whilst it is running.

(4) Do not touch the chips and swarf as they leave the tool point. If chips have to be removed use a brush or stick.

(5) Before starting the machine make sure that the work and tool are properly held.

(6) Always use any guards which are provided.

(7) Stop the machine before attempting to take measurements.

(8) Wear goggles when cutting materials such as cast iron and brass which produce small chips.

Exercise 4.15

1. Sketch two types of guideways used on lathes.
2. What are the two main uses of the tailstock?
3. On the calibrated dial of the cross-slide it is found that 10 revolutions of the handwheel make the cross-slide move 50 mm. The dial has 100 divisions. Find: (a) the distance moved by the cross-slide for each division on the dial; (b) the number of divisions needed to give the cross-slide a movement of 0·035 mm.
4. Each division of the calibrated dial of a cross-slide gives a movement of 0·01 mm. If the reading on the dial is 32 and a cross-slide movement of 0·62 mm is needed, what will be the reading on the dial when this movement is achieved?
5. What are the advantages of the four-way tool post?
6. Sketch an ordinary clamp-type tool post and show on it how the tool is prevented from moving in any one of the six possible ways.
7. How is a flat surface produced on the lathe?
8. How is round work produced on the lathe?
9. When would a 3-jaw chuck be used and when would a 4-jaw type be used?
10. Sketch a round-nosed roughing tool. Mark on your sketch the rake and clearance angles.
11. Why must two cuts always be taken when turning?
12. Why is a finishing cut taken with a finer feed than a roughing cut?
13. Sketch a boring tool marking the rake and clearance angles.
14. Sketch a plain mandrel and state when it would be used.
15. Show on a sketch how the work is prevented from moving in a 3-jaw chuck.
16. What advantages does an expanding mandrel have over a plain mandrel?
17. Describe the procedure of drilling and reaming on a centre lathe.
18. By making clear sketches show the effect on the rake and clearance angles of: (a) setting a tool below the centre-line of the work; (b) setting a tool above the centre of the work.
19. State three reasons for using cutting fluids.
20. Name four kinds of cutting fluids.
21. State the cutting fluids usually used when machining: (a) mild steel; (b) aluminium; (c) bronze; (d) cast iron.
22. Why are boring bars often used in preference to other types of boring tools?
23. Show on clear diagram how the work is prevented from moving when using a faceplate.
24. What is a balance weight and when would it be used?
25. When screw cutting how is the form of the thread obtained? How is the pitch obtained?
26. When using the angular feed method of screw cutting, how is the depth of thread obtained?
27. Show how the tool should be set when cutting an internal thread. To what diameter is the work bored?
28. Why must a screw thread tool be set absolutely square to the work?
29. Why is a chaser used?
30. Calculate suitable change-wheels for cutting the following threads on a lathe with a 6 mm pitch leadscrew: (a) 2 mm (b) 3 mm (c) 4 mm (d) 1·75 mm (e) 2·5 mm (f) 2·25 mm (g) 3·5 mm.

4.16   The shaping and slotting machines

The shaping machine is used for producing flat surfaces. It is more compact and cheaper than a milling machine and it is often speedier, particularly when operating on small work. The shaping machine is frequently used for toughing-out castings since sand or hard spots may damage the teeth of a milling cutter.

The main parts of the shaping machine. These are illustrated in Fig. 4.205:

The shaping-machine driving-mechanism is shown in Fig. 4.206.

The slotted link converts the circular motion of the bull gear into the straight line motion of the ram. Fig. 4.207 shows that the cutting stroke takes longer than the return stroke, and the mechanism is described as a *quick-return* mechanism.

Action of the shaping machine. Fig. 4.208 shows how a flat surface is produced on the shaping machine. It will be seen that the tool movement and the work movement

are perpendicular to each other. The work is moved a short distance during the return stroke and the amount of work movement is called the *feed*. This is usually quoted in mm per stroke (e.g. 0·25 mm per stroke).

In setting for horizontal work the length of the stroke should be regulated so that it is about 20 mm longer than the work. About 12 mm of this additional length should be at the start of the cut. When machining a rectangular block it is more economical in time if the block is set so that the tool moves parallel to the longest side because less time is taken during the return strokes.

The feed mechanism. The feed of the shaper table and the way in which it is obtained varies with different makes of machines. It is not so important to have such a large selection of feeds as it is with the centre lathe and milling machine. The feed is selected according to the finish required—a very coarse feed being used for roughing cuts and a fine feed being used for finishing cuts. Fig. 4.209 shows the feed mechanism.

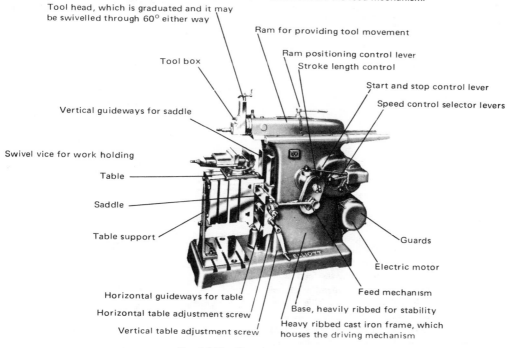

Tool head, which is graduated and it may be swivelled through 60° either way

Ram for providing tool movement

Tool box

Ram positioning control lever
Stroke length control

Start and stop control lever

Speed control selector levers

Vertical guideways for saddle

Swivel vice for work holding

Table

Saddle

Table support

Guards

Electric motor

Horizontal guideways for table

Horizontal table adjustment screw

Vertical table adjustment screw

Feed mechanism

Base, heavily ribbed for stability

Heavy ribbed cast iron frame, which houses the driving mechanism

Fig. 4.205   The shaping machine   *(B. Elliott (Machinery) Ltd.)*

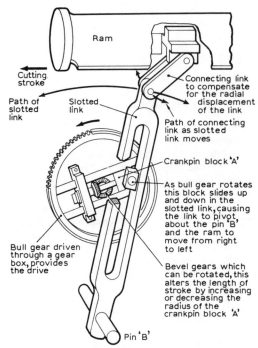

Fig. 4.206   The shaping-machine driving-mechanism

Ram

Cutting stroke

Path of slotted link

Slotted link

Connecting link to compensate for the radial displacement of the link

Path of connecting link as slotted link moves

Crankpin block 'A'

As bull gear rotates this block slides up and down in the slotted link, causing the link to pivot about the pin 'B' and the ram to move from right to left

Bull gear driven through a gear box, provides the drive

Bevel gears which can be rotated, this alters the length of stroke by increasing or decreasing the radius of the crankpin block 'A'

Pin 'B'

Cutting takes place as the slotted link moves through this angle

Extreme forward position of the slotted link

Extreme backward position of the slotted link

Return stroke takes place as the slotted link moves through this angle

Fig. 4.207

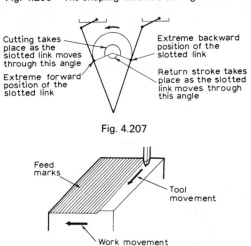

Feed marks

Tool movement

Work movement

Fig. 4.208   The action of a shaping machine

The table feed depends upon the number of teeth on the ratchet wheel and the pitch of the thread on the feed shaft. If the ratchet wheel has 40 teeth and the feed-shaft thread has a pitch of 6 mm, then 1 revolution of the ratchet wheel moves the feed shaft through 6 mm. This means that if the pawl moves one tooth at a time the table feed will be $\frac{6}{40}$ = 0·15 mm per stroke. If the feed is set so that the pawl moves three teeth at a time, the table feed will be 3×0·15 = 0·45 mm per stroke.

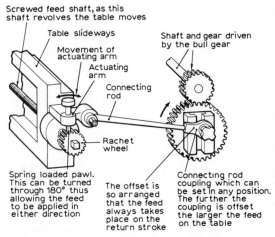

Screwed feed shaft, as this shaft revolves the table moves

Table slideways

Movement of actuating arm

Actuating arm

Connecting rod

Shaft and gear driven by the bull gear

Rachet wheel

Spring loaded pawl. This can be turned through 180° thus allowing the feed to be applied in either direction

The offset is so arranged that the feed always takes place on the return stroke

Connecting rod coupling which can be set in any position. The further the coupling is offset the larger the feed on the table

Fig. 4.209   The shaping-machine feed-mechanism

Cutting speeds. The shaping machine has a reciprocating motion and the tool travels at different speeds for every part of the ram movement. The cutting speed obtained depends on both the gear-box setting and the length of stroke. With a particular gear-box setting the cutting speed alters if the length of stroke is changed. The cutting speeds quoted are always the average cutting speeds which can be calculated from the formula:

$$average\ cutting\ speed\ in\ metres\ per\ minute$$
$$= \frac{length\ of\ stroke\ in\ metres}{time\ in\ minutes\ taken\ by\ the\ cutting\ stroke}$$

Example. During a shaping operation the length of the cutting stroke is 500 mm and the machine makes 30 cutting strokes per minute. Find the average cutting speed.

Length of stroke = 500 mm = 0·5 m

Time taken for 1 cutting stroke = $\frac{1}{30}$ min

Average cutting speed = $0·5 \div \frac{1}{30} = 0·5 \times \frac{30}{1} = 15 \text{m/min}$

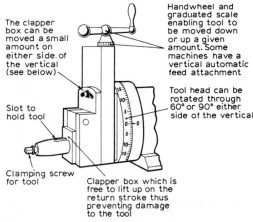

Handwheel and graduated scale enabling tool to be moved down or up a given amount. Some machines have a vertical automatic feed attachment

The clapper box can be moved a small amount on either side of the vertical (see below)

Slot to hold tool

Clamping screw for tool

Tool head can be rotated through 60° or 90° either side of the vertical

Clapper box which is free to lift up on the return stroke thus preventing damage to the tool

Fig. 4.210   The sliding head of the shaping machine

Holding the tool. Fig. 4.210 illustrates the sliding head of the shaping machine.

Fig. 4.211 shows how the toolhead and clapper box is set when shaping a horizontal surface. This position prevents the tool from digging in and also enables the tool to swing away from the work on the return stroke.

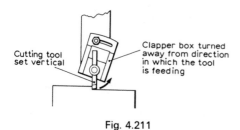

Cutting tool set vertical

Clapper box turned away from direction in which the tool is feeding

Fig. 4.211

Holding the work. Except for castings and work of irregular shape, the machine vice is used for holding the work. The jaw pieces of the vice are usually left soft and hence they can be cleaned up if they become damaged. Castings, etc., can be clamped to the table in the usual way by using bolts and setting pieces.

Fig. 4.212 shows the work being held in the machine vice. Notice how the main cutting force is resisted by the fixed jaw of the vice.

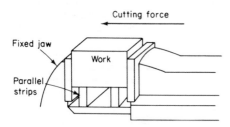

Cutting force

Fixed jaw

Work

Parallel strips

Fig. 4.212   Work held in the machine vice

Machining vertical and angular faces. The methods are shown in Figs 4.213 and 4.214.

Feed

Fig. 4.213   Machining a vertical face

Feed

Fig. 4.214   Machining an angular face

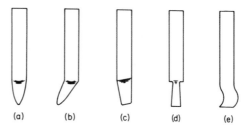

(a) (b) (c) (d) (e)

(a) Round-nosed roughing tool
(b) Cranked tool
(c) Straight-nosed roughing tool
(d) Slotting tool
(e) Swan-necked finishing tool

Fig. 4.215   Shaping machine tools

Shaping tools. The tools used in a shaping machine are similar to those used on a lathe but the shanks are generally stronger so that they can withstand the shock which occurs at the beginning of each cutting stroke. Fig. 4.215 shows a selection of shaping tools.

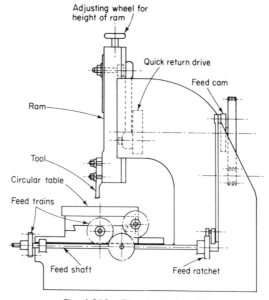

Adjusting wheel for height of ram

Quick return drive

Feed cam

Ram

Tool

Circular table

Feed trains

Feed shaft

Feed ratchet

Fig. 4.216   The slotting machine

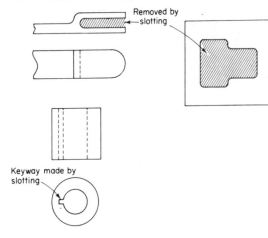

Keyway made by slotting

Fig. 4.217   Typical slotting work

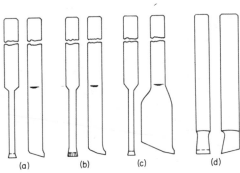

(a)   (b)   (c)   (d)

(a) Keyway tool
(b) Round-nosed tool
(c) Parting tool
(d) Relieving tool

Fig. 4.218   Slotting machine tools

The slotting machine. This machine is, in effect, a vertical shaping machine. It is used for machining the internal surfaces of forgings and castings and for profiling large plates. The main parts of the slotting machine are shown in Fig. 4.216.

Some typical work done on slotting machines is shown in Fig. 4.217.

A selection of the tools used on a slotting machine are shown in Fig. 4.218.

Safety

(1) Before starting the machine make sure that the work and tool are securely held.

(2) Always stop the machine before taking any measurements.

(3) Wear goggles when cutting brittle materials like brass and cast iron.

Exercise 4.16

1. Show on clear sketches how the clapper box should be set when machining: (a) a horizontal surface; (b) a vertical surface; (c) an angular surface.

2. Why is the slotted-link mechanism used on a shaping machine?

3. The feeding mechanism of a shaping machine has a ratchet wheel which has 40 teeth and a feed shaft with a 5 mm pitch thread. If the feed is set so that the pawl moves two teeth at a time, what is the table feed?

4. The stroke of a shaping machine is set at 40 mm and the ram makes 30 cycles per minute. What is the average cutting speed?

5. Why are shaping tools more robust than lathe tools?

6. Sketch a shaping-machine tool of your choice and mark on it the rake and clearance angles.

7. When holding work in the machine vice show how the main cutting force is resisted.

8. State three kinds of work which would be done on a slotting machine.

9. Sketch a round-nosed slotting-machine tool and mark on it the rake and clearance angles.

4.17 The milling machine

Milling machines are extremely versatile. It is possible to produce plane and cylindrical surfaces on these machines as well as a variety of other shapes. Milling will produce flat surfaces more quickly and accurately than the shaping machine and hence milling machines are used extensively for production work.

Types of milling machine. The main types are:
(1) the horizontal machine (or plain milling machine);
(2) the vertical machine;
(3) The universal machine.
   Each of these machines is described below.

1. *Horizontal milling machine*, Fig. 4.219:
2. *The universal machine* is similar to the horizontal machine but it has a table which swivels to allow helixes, etc., to be milled.
3. *Vertical milling machine*, Fig. 4.220:

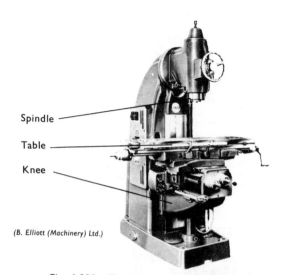

Spindle

Table

Knee

(B. Elliott (Machinery) Ltd.)

Fig. 4.220   The universal milling machine

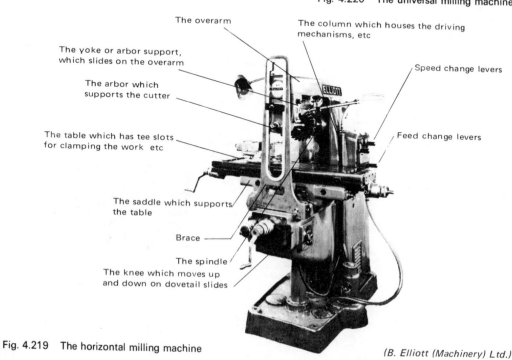

The overarm

The yoke or arbor support, which slides on the overarm

The arbor which supports the cutter

The table which has tee slots for clamping the work etc

The saddle which supports the table

Brace

The spindle

The knee which moves up and down on dovetail slides

The column which houses the driving mechanisms, etc

Speed change levers

Feed change levers

Fig. 4.219   The horizontal milling machine

(B. Elliott (Machinery) Ltd.)

Care and lubrication of milling machines. The table must be kept clean and it must not be damaged when clamping work on to it. The slides must be oiled regularly and kept clean. Modern machines are fitted with an automatic lubricating system which ensures a good supply of lubricant to all of the important parts. The yoke bearings must be oiled frequently.

Action of the horizontal milling machine. The workpiece is held in a machine vice or a special fixture or it is clamped to the table. The cutting is performed by feeding the work under a revolving cutter. The table movements possible are shown in Fig. 4.221.

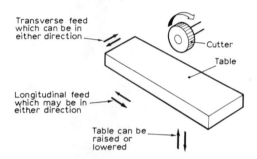

Fig. 4.221   Action of a horizontal milling machine

Two methods of horizontal milling are possible, as shown in Figs 4.222 and 4.223.

1. Upcut or conventional milling, Fig. 4.222:

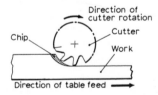

Fig. 4.222   Upcut milling

2. Downcut or climb milling, Fig. 4.223. This method has the advantage that the teeth cut downwards and hence there is no tendency to lift the work. It should, however, be avoided unless the machine is fitted with a special feed screw and nut to avoid backlash.

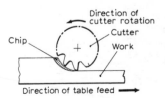

Fig. 4.223   Downcut milling

Action of the vertical milling machine. The table movements and the up and down movement of the spindle are shown in Fig. 4.224. The vertical machine can produce horizontal surfaces or vertical surfaces.

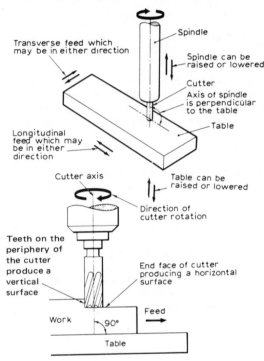

Fig. 4.224   Action of a vertical milling machine

Milling cutters. The commonest types of milling cutter are illustrated below.

1. Cylindrical cutter (slab mill or roller mill), Fig. 4.225. This cutter is used for producing horizontal surfaces and can be obtained in a wide range of diameters and widths. Cutters with a large number of teeth are used for light work, those with fewer teeth being used for heavy work.

(A. A. Jones & Shipman Ltd.)

Fig. 4.225   Cylindrical cutter

2. Side and face cutter, Fig. 4.226. This cutter has teeth cut on both its sides as well as on its periphery. It is used to produce slots and, when used in pairs, to produce flats, squares, hexagons, etc. In the larger sizes, the teeth are made separately and are inserted into the body of the cutter. This is advantageous since tool inserts can be removed and replaced if they are broken.

(A. A. Jones & Shipman Ltd.)

Fig. 4.226   Side and face cutter

3. *Slotting cutter*, Fig. 4.227. This cutter has teeth only on the periphery and it is used for cutting slots and keyways.

(A. A. Jones & Shipman Ltd.)

Fig. 4.227   Slotting cutter

4. *Metal-slitting saw*, Fig. 4.228. This cutter may have teeth on the periphery only or it may have teeth both on the periphery and sides. It is used to cut deep slots and to cut material to length. The thickness of the cutter varies in thickness from 1 mm to 5 mm and it is thinner at the centre than at the edge to prevent the cutter from binding in the slot.

(A. A. Jones & Shipman Ltd.)

Fig. 4.228   Metal-slitting saw

5. *End mills*. End mills are used from about 4 mm diameter upwards. Cutters up to about 40 mm diameter have

the cutter and shank in one piece and as shown in Fig. 4.229 the shank may be either parallel or tapered.

Parallel shank held in special collet.

Teeth are provided on the end face and on the periphery.

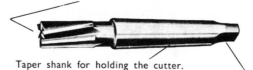

Taper shank for holding the cutter.

Tang for removing the cutter.

Fig. 4.229   End mills

6. *Shell end mills*, Fig. 4.230. Shell end mills are made to be fitted on to stub arbors (see the following section) which take the place of the shanks. They are thus cheaper to replace than solid end mills.

Fig. 4.230   Shell end mill

7. *Face mill*, Fig. 4.231. This cutter is designed to take heavy cuts and it is used to produce flat surfaces. It is more accurate than the cylindrical slab mill. The larger kinds have inserted teeth. Face mills have teeth on the end face and on the periphery. The length of the teeth on the periphery is always less than half the diameter of the cutter.

Fig. 4.231   Face mill

(A. A. Jones & Shipman Ltd.)

Fig. 4.232   Tee-slot cutter

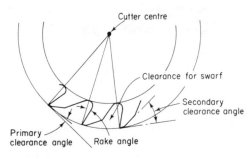

Fig. 4.233   Cutting angles for a milling cutter

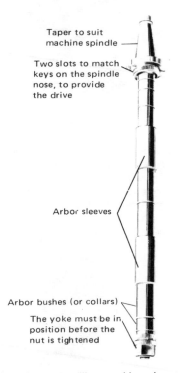

Fig. 4.234   A milling-machine arbor

8. *Tee-slot cutter*, Fig. 4.232. This cutter is used for milling tee slots. A slot or a groove must be made in the work before this cutter can be used.

The form of a milling-cutter tooth is shown in Fig. 4.233. The multi-toothed cutter requires a secondary clearance angle as shown in the diagram. The rake angle is measured relative to a radial line, as shown.

When using a straight-toothed milling-cutter one tooth completes its cut before the next tooth comes into action. A break in the cutting therefore occurs. When helical cutters are used this break in the cutting is avoided and hence the cutting action is much smoother. The cutting forces on the teeth of the helical cutters are smaller than those on the straight-toothed cutters, which means that less wear occurs on the teeth of the helical cutters.

Mounting the cutters. On horizontal machines the cutters are always carried on an arbor such as the one illustrated in Fig. 4.234.

It is important for the arbor to run true. Unless the arbor runs true the cutter cannot run true. Therefore bending of the arbor should be reduced as much as possible by the methods shown in Fig. 4.235.

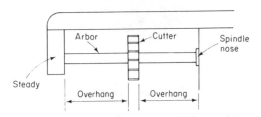

The excessive overhang causes a large deflection of the arbor causing the cutter to run out of true. Chattering and poor surface finish on the work will result

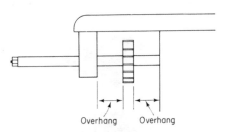

By reducing the overhang to a minimum the arbor deflects only a very small amount and consequently the cutter runs true

Fig. 4.235   Effect of overhang when mounting cutters

On small machines the cutter is driven by friction between the spacing bushes (or collars) and the cutter. On larger machines the cutter is keyed to the arbor.

Spacing collars

The collars are hardened and ground and the end faces are parallel to each other and square to the bore

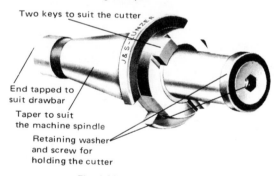

(A. A. Jones & Shipman Ltd.)

Fig. 4.236   Spacing collars

The *stub arbor*, Fig. 4.237, is generally used on the vertical machine when using face milling cutters and shell end mills. It can also be used on the horizontal machine when facing is required.

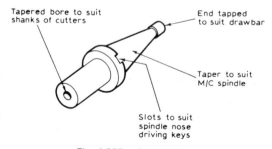

Two keys to suit the cutter

End tapped to suit drawbar

Taper to suit the machine spindle

Retaining washer and screw for holding the cutter

Fig. 4.237   The stub arbor

The *cutter adaptor*, Fig. 4.238, is used to hold small cutters with tapered shanks.

Tapered bore to suit shanks of cutters

End tapped to suit drawbar

Taper to suit M/C spindle

Slots to suit spindle nose driving keys

Fig. 4.238   Cutter adaptor

Cutting speeds. The table below gives suggested cutting speeds for high-speed-steel cutters when taking roughing cuts. Finishing speeds can be from 25% to 100% higher than those suggested for roughing.

CUTTING SPEEDS

Material to be cut	Cutting speed metres/min
Mild steel	20
High-carbon steel	12
Cast iron	15
Aluminium	90–180
Brass	30–45
Bronze	12–23
Copper	45

Feeds for milling cutters. Feeds for milling cutters are usually expressed in millimetres per tooth since this gives a good guide to the work that each tooth does in cutting. The feed rate must not be too great because excessive load on the teeth can cause cutter breakage, overheating or clogging of the cutter. The table below will give a guide for feeds which may be used with various materials and different types of cutters.

FEEDS FOR MILLING CUTTERS

Material to be cut	Feed in millimetres per tooth			
	Slab mills	Side mills Slotting cutters	Saws	End mills
Mild steel	0·23	0·18	0·08	0·13
Tool steel	0·18	0·15	0·08	0·13
Cast iron	0·30	0·23	0·10	0·20
Aluminium	0·43	0·33	0·13	0·28
Medium bronze	0·36	0·28	0·10	0·23

The feed with any kind of cutter should be as high as possible. The values given in the table should be scaled down where the depth of cut is large and where the work or cutter is fragile.

The feed on the milling machine table is usually stated in millimetres per minute. The method of converting a feed in millimetres per tooth to a feed in millimetres per minute is given in Chapter 1.11.

Work-holding devices. The most common way of holding the work is by using the swivel vice (Fig. 4.239). Large components are clamped on to the machine table.

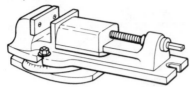

Fig. 4.239 The swivel vice. Wherever possible the cutting forces should be directed so that the fixed jaw acts as the restraint

When, for instance, a series of flats is to be milled on a bar an indexing device is used. One such device is the rotary table (Fig. 4.240), which may be fitted with a collet or a chuck to hold the work (Fig. 4.241). It is suitable for the direct indexing of 2, 3, 4, 6, 8, 12 and 24 divisions, but as it is also graduated in degrees it can be locked in any position.

Fig. 4.240 A rotary table

Fig. 4.241 Rotary table fitted with (a) a collet (b) a chuck

Safety. The milling machine is the most dangerous machine tool of all.

(1) Guards must always be in position when cutting.

(2) Do not remove swarf when the machine is running.

(3) Stop the machine before taking measurements.

Exercise 4.17

1. Name three kinds of milling machines.

2. By using clear sketches show the differences between upcut and downcut milling.

3. Describe the main features of the following milling cutters: (a) cylindrical; (b) side and face; (c) slitting saw; (d) face mill; (e) shell end mill.

4. When is a stub arbor used?

5. When is a cutter adaptor used?

6. Why is the feed on a milling machine stated in millimetres per tooth?

7. Show how the rake and clearance angles are produced on a straight-toothed milling-cutter.

8. When mounting cutters on the arbor why should excessive overhang be avoided?

9. Show how a milling cutter may be held on the arbor by friction.

10. When using a swivel vice, why must the cutting force be directed towards the fixed jaw?

11. When should a rotary table be used?

4.18   Electrical installation

Safety. The main dangers when using electricity are shock, burns, fire and explosion. The Institution of Electrical Engineers (I.E.E.) have laid down standards in their wiring regulations which must be adhered to in all circumstances.

Shock arises when the metal framework of an appliance becomes live due to damaged wiring. Sustained over-loading of equipment can lead to overheating with consequent fire risk. Some of the ways in which these hazards are overcome are discussed later in this Chapter.

It is essential that all conductors and equipment are sufficient in size and power for the work that they are called upon to do. They must be installed in such a way that danger is prevented as far as possible.

Cables. Cables consist of two parts:
(1) a conductor which carries the current;
(2) insulation which confines the current to the conductor. Most conductors are made of either copper or aluminium. The main insulating materials are vulcanized rubber, PVC, polythene and impregnated paper. Some cables have a rubber-insulated conductor enclosed in a lead or aluminium sheath. However, cables insulated and sheathed with PVC have almost entirely replaced the metal-sheathed cables. Some typical cables are shown in Fig. 4.242.

The flexibility of a cable depends on the number of strands making up the conductor. Portable equipment such as soldering irons, drills, etc., are usually connected to a supply by a many-stranded cable. The size of a cable chosen for a particular application depends upon the current it has to carry. If the cable is overloaded it will overheat and hence an adequate size of cable must be chosen.

When removing the outer sheathing of a cable care must be taken to avoid cutting through the insulation covering the conducting wires. The outer sheathing may be removed by using a penknife or similar tool. Sufficient of the insulation around the conducting wires must then be removed so that a connection can be made. Special pliers are used for this purpose. These are made so that they can be adjusted to cut only the insulation without fear of cutting the conducting wires.

Conduit. Steel tubes or conduits are used to protect cables. Two types are available:

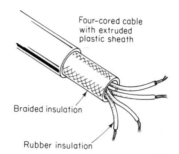

Four-cored cable
with extruded
plastic sheath

Braided insulation

Rubber insulation

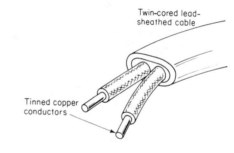

Twin-cored lead-
sheathed cable

Tinned copper
conductors

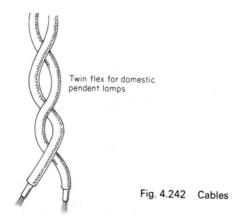

Twin flex for domestic
pendent lamps

Fig. 4.242   Cables

1. *Heavy-gauge screwed conduit* in which all the joints are made by screw threads and unions.
2. *Light-gauge conduit* in which the joints are made by means of grip sockets and fittings (Fig. 4.243).

Conduit is obtained in sizes ranging from 15 mm inside diameter to 50 mm inside diameter. The size of

Grip inspection
box

Normal bend (screwed)

Inspection elbow
(screwed)

Fig. 4.243   Conduit fittings

the conduit used depends upon the number and size of cables which the conduit is to carry. The maximum number of cables permitted do not fill the conduit since an air space is needed to prevent overheating.

Cable connections. When wires are joined together or connected to the terminals of an appliance the joint or connection must possess both electrical continuity and adequate mechanical strength. The methods used depend upon the currents to be carried. Some typical connections and joints are shown in Fig. 4.244.

Protection of circuits. In every electrical circuit there is a danger that the circuit will be overloaded. A fuse is the simplest way of giving protection to the circuit when the current becomes too great.

When electricity flows through a conductor the conductor increases in temperature. If the resistance of the conductor is high there will be a rapid rise in temperature which can cause the conductor to melt. A wire of tinned

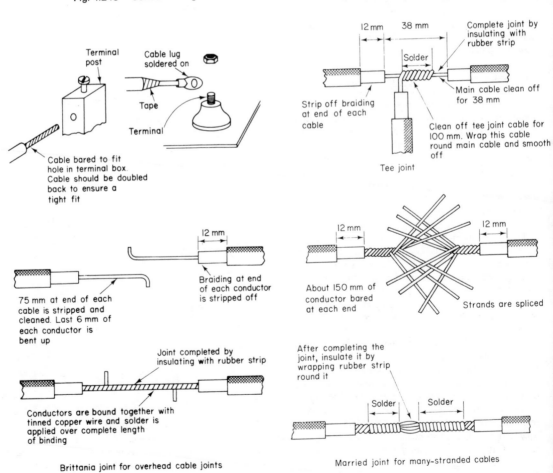

Fig. 4.244   Cable connections and joints

copper or silver is placed in the circuit as a safety device. When the current reaches a value nearing the safe limit for the circuit wiring the piece of wire melts, so breaking the circuit. A wire used in this way is called a *fuse*.

A rewirable fuse (Fig. 4.245) is simple and cheap but it has the following disadvantages:

(1) It is easy for an inexperienced person to replace the fuse with wire of the wrong thickness and material.

(2) The fuse will often deteriorate due to rusting.

Cartridge fuses (Fig. 4.246) are better and many fuse carriers are designed so that only the correct cartridge fuse can be fitted in them. It is then impossible to use the incorrect fuse.

Both the rewirable and cartridge types are used in industrial installations but for large installations the high rupturing capacity (H.R.C.) fuse is used. This type is more expensive than the other types.

Earthing. The purpose of earthing is to make sure that whenever a live wire comes into contact with the metal frame of an electrical appliance the current has an easy path to earth. When this happens the fuse blows and the current is cut off. If the appliance is not earthed the metal frame will become live with consequent danger of shock to personnel and/or fire risk.

Earthing is done by making a direct connection between the appliance and a metal water main or an earth plate. An earth plate is a plate buried in the ground so that it has good electrical connection with the surrounding earth.

It is absolutely essential that portable power tools and similar equipment are properly earthed. The earthing is usually made through the largest pin of a three-pin plug as shown in Fig. 4.247.

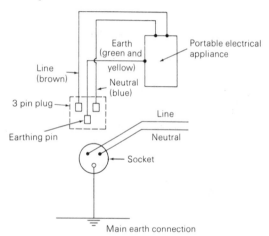

Fig. 4.247 Earthing through the largest pin of a three-pin plug

Fig. 4.245 Rewirable fuses and fuse wires

The colour codes for cartridge fuses are
2A – yellow
5A – red
10A – black
13A – brown
15A – blue

Fig. 4.246 Cartridge fuse

Exercise 4.18

1. Name the three main risks associated with electrical power.
2. Name three insulating materials used for cables and name two commonly used conducting metals.
3. What precautions should be taken when stripping cables?
4. What is the principle of the fuse?
5. Why is portable equipment earthed?
6. Name three cable joints stating the use of each.
7. What does the flexibility of a cable depend on?
8. For what purpose is conduit used? State two kinds.
9. Why do the cables not completely fill conduit?
10. How is portable equipment earthed?

Revision questions

Each question contains four alternative answers marked a, b, 'c and d. Only ONE answer is correct. In answering each question ring the letter which corresponds to the answer you think is correct.

1. What instrument would be most appropriate for measuring an external diameter nominally 25 mm diameter to an accuracy of 0·05 mm?
   (a) outside caliper
   (b) adjustable-caliper limit-gauge
   (c) micrometer caliper
   (d) vernier caliper

2. A metal which can easily be drawn into thin wire must possess good
   (a) elasticity   (b) malleability
   (c) ductility   (d) hardness

3. An example of a non-combustible gas is
   (a) propane   (b) acetylene
   (c) hydrogen   (d) oxygen

4. The total resistance of the parallel circuit (Fig. 1) is
   (a) less than 0·5 Ω   (b) 50 Ω
   (c) 100·0 Ω   (d) greater than 100·5 Ω

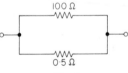

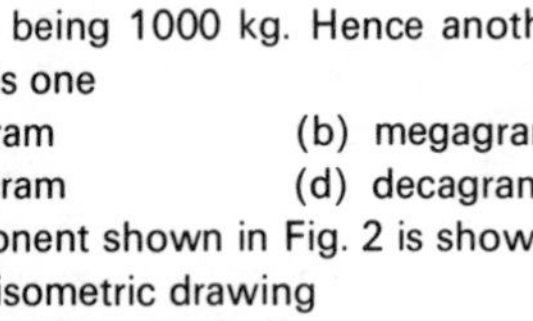

Fig. 1

5. A commonly used metric unit of mass is the tonne, one tonne being 1000 kg. Hence another name for the tonne is one
   (a) milligram   (b) megagram
   (c) centigram   (d) decagram

6. The component shown in Fig. 2 is shown
   (a) as an isometric drawing
   (b) as an oblique drawing
   (c) in first-angle projection
   (d) in third-angle projection

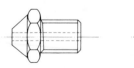

Fig. 2

7. A fine-tooth hacksaw blade having 10 teeth per centimetre is most suitable for cutting
   (a) large-section mild steel
   (b) aluminium bar
   (c) thin-walled tubing
   (d) cast iron

8. A treadle guillotine is mainly used for
   (a) shearing sheet metal
   (b) folding sheet metal
   (c) nibbling
   (d) wiring edges

9. Which of the sketches in Fig. 3 shows a bolt?

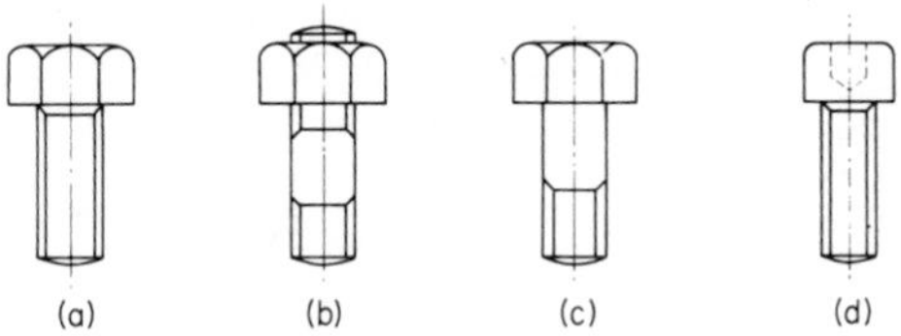

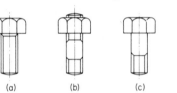

(a)   (b)   (c)   (d)

Fig. 3

10. The included angle of the tapered point in Fig. 4 is
    (a) 180° − A   (b) ½ A
    (c) A   (d) 2 A

11. The centre shown in Fig. 4 is called
    (a) running centre   (b) live centre
    (c) half centre   (d) dead centre

Fig. 4

12. Fig. 5 shows a lathe cutting tool set on the centre of rotation. The angle A is the
    (a) point angle   (b) wedge angle
    (c) rake angle   (d) clearance angle

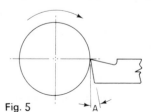

Fig. 5

13. The thread angle of a metric screw is
    (a) 29°   (b) 47½°
    (c) 55°   (d) 60°
14. The tenon of a milling fixture is to fit into the tee slot on a milling-machine table. If the limits of size are as shown in Fig. 6, the minimum clearance will be
    (a) nil   (b) 0·01 mm
    (c) 0·02 mm   (d) 0·03 mm

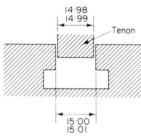

All dimensions in millimetres

Fig. 6   All dimensions in millimetres

15. The symbol which indicates a fillet weld is

    (a)   (b)
    (c)   (d)

16. Pipe installations are colour coded to indicate the
    (a) substance flowing through the pipe
    (b) material from which the pipe is made
    (c) internal diameter of the pipe
    (d) temperature of the contents
17. 0·125° is
    (a) 0° 12′ 5″   (b) 0° 7′ 5″
    (c) 0° 7′ 30″   (d) 0° 7′ 50″
18. Set squares are placed together as shown in Fig. 7. The angle θ is
    (a) 5°   (b) 15°
    (c) 45°   (d) 75°

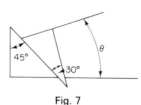

Fig. 7

19. Odd-legs are suitable for
    (a) marking off the centre of a hole which is to be drilled
    (b) marking off a line parallel to the surface table
    (c) marking off an angle
    (d) marking off a line at a fixed distance from an edge of a plate

20. A twist drill is ground so that one lip is longer than the other. This will result in the drill
    (a) rubbing instead of cutting
    (b) breaking
    (c) cutting an oversize hole
    (d) burning at the cutting edge
21. Which screw (Fig. 8) has a cheese head?

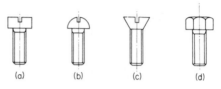

    (a)   (b)   (c)   (d)

Fig. 8

22. The joint shown in Fig. 9 is called a
    (a) butt joint   (b) toggle joint
    (c) lap joint   (d) seamed joint

Fig. 9

23a. Fig. 10 shows a
    (a) gap gauge   (b) depth gauge
    (c) plug gauge   (d) profile gauge
23b. The tolerance on the part to be checked by the gauge in Fig. 10 is
    (a) 19·05 mm   (b) 0·14 mm
    (c) 18·92 mm   (d) ±0·07 mm

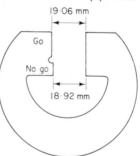

Fig. 10

24. Fig. 11 shows a cold chisel in the act of cutting. The angle marked A is the
    (a) clearance angle   (b) rake angle
    (c) cutting angle   (d) point angle

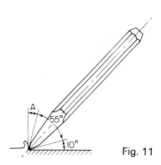

Fig. 11

25. The angle marked A in Fig. 11 is
    (a) 35°   (b) 37° 30′
    (c) 45°   (d) 65°
26. The lathe tool shown in Fig. 12 is a
    (a) knife tool   (b) parting-off tool
    (c) boring tool   (d) finishing tool

Fig. 12

27. Which of the following is usually machined without a coolant?
    (a) mild steel   (b) cast iron
    (c) aluminium   (d) high-carbon steel
28. A mandrel is used in turning to make sure that
    (a) the bore is not tapered
    (b) the bore is the correct size
    (c) the outside diameter is concentric with the bore
    (d) the outside diameter is the correct size
29. Portable electrical equipment is earthed
    (a) to increase the efficiency of the equipment
    (b) to increase the resistance of the equipment
    (c) to reduce the current flowing
    (d) to reduce electrical danger
30. The edges of sheet metal are folded or wired
    (a) to prevent corrosion
    (b) to help the solder to run
    (c) to make the edge stronger and safe
    (d) to enable the edge to be grasped in a vice
31. After a chisel has been forged to shape, hardened and tempered, the cutting end is found to be cracked. This could be due to
    (a) forging when the steel is below red heat
    (b) hardening the chisel from below red heat
    (c) tempering the chisel from too high a temperature
    (d) not quenching the chisel after tempering
32. Which of the symbols shown in Fig. 13 indicates an electrical fuse.

Fig. 13

33. Which of the beaded edges in Fig. 14 gives the strongest edge?

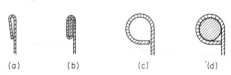

Fig. 14

34. The rectangular block shown in Fig. 15 is to be machined on the top face. If a shaping machine is used the most practical economic length of stroke would be
    (a) 40 mm   (b) 75 mm
    (c) 95 mm   (d) 50 mm

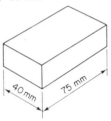

Fig. 15

35. When dowels are used in assembly work their purpose would be
    (a) to make a tighter joint
    (b) to reduce the number of bolts required
    (c) to make the use of nuts unnecessary
    (d) to position the components accurately
36. A half-round scraper is suitable for
    (a) scraping flat surfaces
    (b) scraping bearings
    (c) removing burrs
    (d) scraping keyways
37. The cutting edge of a cold chisel needs to be
    (a) case hardened
    (b) hardened and tempered
    (c) annealed
    (d) normalized
38. When steel is worked when cold it
    (a) recrystallizes
    (b) becomes softer
    (c) work hardens
    (d) becomes malleable
39. The locking of a nut depends upon either frictional or positive locking. Which of the following depends upon friction?
    (a) tab washer
    (b) split pin
    (c) locking wire
    (d) fibre-insert lock-nut
40. The reading on the micrometer shown in Fig. 16 is
    (a) 7·04 mm   (b) 70·4 mm
    (c) 7·06 mm   (d) 7·60 mm

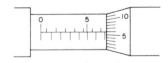

Fig. 16

41. The reading of the vernier caliper shown in Fig. 17 is
    (a) 28·72 mm
    (b) 25·60 mm
    (c) 2·56 mm
    (d) 75·60 mm

Fig. 17

42. The tolerance on the dimensions positioning a hole are ±0·04 mm. The hole centre should be marked off by using
    (a) inside calipers and rule
    (b) odd-leg calipers and rule
    (c) a try-square
    (d) a bevel

43. The number 7507 correct to 2 significant figures is
    (a) 75
    (b) 751
    (c) 750
    (d) 7500

44. 1 kilometer is equal to
    (a) 1000 mm
    (b) 100 cm
    (c) 10 000 cm
    (d) 1 000 000 mm

45. Which of the following fractions is equal to $\frac{4}{9}$?
    (a) $\frac{15}{27}$
    (b) $\frac{4}{36}$
    (c) $\frac{36}{4}$
    (d) $\frac{52}{117}$

46. When $6\frac{4}{9}$ is divided by $3\frac{2}{3}$ the answer is
    (a) $2\frac{2}{3}$
    (b) $\frac{638}{27}$
    (c) $\frac{58}{33}$
    (d) $18\frac{8}{27}$

47. The number 0·075 538 correct to two places of decimals is
    (a) 0·076
    (b) 0·075
    (c) 0·07
    (d) 0·08

48. A line 920 cm long is divided into four parts in the ratios 15:13:10:8. The longest part is
    (a) 260 cm
    (b) 200 cm
    (c) 300 cm
    (d) 160 cm

49. A quantity of alloy has a mass of 400 kg. It contains copper, lead and tin in the ratios by mass of 15:3:2. The mass of lead in the alloy is
    (a) 300 kg
    (b) 60 kg
    (c) 40 kg
    (d) 200 kg

50. 30% of a certain length is 600 mm. The complete length is
    (a) 20 mm
    (b) 200 mm
    (c) 2 m
    (d) 630 mm

51. The average of three numbers is 116. The average of two of them is 98. The third number is
    (a) 18
    (b) 107
    (c) 110
    (d) 152

52. The average length of 48 bars of metal is 138 cm. 8 bars are removed and the average length of the remaining bars is 135 cm. The average length of the 8 bars which were removed is
    (a) 160 cm
    (b) 159 cm
    (c) 132 cm
    (d) 156 cm

53. The square root of 0·036 is equal to
    (a) 0·6
    (b) 0·06
    (c) 0·1897
    (d) 0·018 97

54. The value of $0·09^2$ is
    (a) 0·81
    (b) 0·081
    (c) 0·0081
    (d) 0·000 81

55. If x and y are two numbers then four times the sum of the two numbers divided by the first number can be written symbolically as
    (a) $\frac{4x+y}{x}$
    (b) $\frac{x+4y}{x}$
    (c) $\frac{4(x+y)}{x}$
    (d) $\frac{x+y}{4x}$

56. If $x = 2$, $y = 3$ and $z = 4$ then the value of xyz^2 is
    (a) 21
    (b) 96
    (c) 576
    (d) 432

57. The solution of the equation $4x - 3 = 5$ is
    (a) $x = 2$
    (b) $x = \frac{1}{2}$
    (c) $x = 8$
    (d) $x = 1$

58. The distance AB (Fig. 18) is required for checking purposes. It is
    (a) 7 cm
    (b) 17 cm
    (c) 13 cm
    (d) 10·91 cm

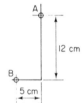

Fig. 18

59. 1 square metre is equal to
    (a) 100 cm²
    (b) 1000 cm²
    (c) 10 000 cm²
    (d) 1 000 000 cm²

60. A measurement is taken as 0·000 008 m. This can be written as
    (a) 8 Mm
    (b) 8 km
    (c) 8 mm
    (d) 8 μm

61. In radio work the term megacycles is used. 1700 megacycles is equal to
    (a) 17 kilomegacycles
    (b) 17 gigacycles
    (c) 17 hectocycles
    (d) 17 decacycles

62. A rectangle is 200 mm long and 150 mm wide. Its area is
    (a) 3 cm²
    (b) 30 cm²
    (c) 300 cm²
    (d) 3000 cm²

63. When turning a bar of mild steel 30 mm diameter the cutting speed is 27 m/min. Hence a suitable spindle speed is
    (a) 150 rev/min
    (b) 200 rev/min
    (c) 220 rev/min
    (d) 270 rev/min

64. A milling cutter 200 mm diameter has 15 teeth. If the cutting speed is to be 40 m/min and the feed per tooth is to be 0·10 mm per tooth the table feed is
    (a) 90 mm/min
    (b) 70 mm/min
    (c) 60 mm/min
    (d) 30 mm/min

65. The graph (Fig. 19) shows the relationship between the spindle speed of a drilling machine and the drill size. The rev/min required by a 5 mm drill is about
    (a) 3000 rev/min   (b) 2000 rev/min
    (c) 1200 rev/min   (d) 800 rev/min

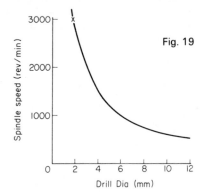

Fig. 19

66. A tough material will
    (a) break when hit with a hammer
    (b) stand repeated blows with a hammer
    (c) withstand a large load without breaking
    (d) easily be marked by a centre punch
67. Copper is used for soldering irons because
    (a) it is a strong metal
    (b) it is a ductile metal
    (c) it has a clean surface
    (d) it is a good conductor of heat
68. High-speed steel has replaced high-carbon steel as a cutting-tool material because
    (a) it wears better
    (b) it is easier to sharpen
    (c) it retains its hardness at quite high temperatures
    (d) it is an alloy steel
69. A thermo-setting plastic
    (a) can be reshaped by applying heat
    (b) can be hardened by applying heat
    (c) cannot be resoftened by applying heat
    (d) can be easily forged
70. Which of the following is a thermo-setting material?
    (a) nylon   (b) polythene
    (c) P.V.C.   (d) bakelite
71. The force shown in Fig. 20 causes the material
    (a) to bend   (b) to twist
    (c) to shear   (d) to stretch

Fig. 20

72. When oxygen is removed from the surface of a metal the process is called
    (a) reduction   (b) oxidation
    (c) corrosion   (d) combustion

73. In oxy-acetylene welding a carburizing flame contains
    (a) no oxygen
    (b) an excess of acetylene
    (c) an excess of oxygen
    (d) equal amounts of oxygen and acetylene
74. Galvanizing is the process of
    (a) painting the steel with a zinc paint
    (b) electroplating the steel to give it a coat of zinc
    (c) spraying the steel with zinc powder
    (d) immersing the steel in a bath of molten zinc
75. Which heat treatment is used to soften steel so that it becomes easier to work in the cold state?
    (a) annealing   (b) normalizing
    (c) tempering   (d) recrystallization
76. An optical pyrometer relies on one of the effects of heat for its operation. The heat effect is
    (a) expansion   (b) change of state
    (c) heat colour   (d) electrical
77. A soldering iron has a mass of $\frac{1}{2}$ kg. The temperature of the iron is 220 °C when soldering commences and after soldering the temperature drops to 150 °C. If the specific heat of copper is 0·4 kJ/kg/°C how much heat energy has been lost by the iron?
    (a) 110 kJ   (b) 75 kJ
    (c) 14 kJ   (d) 70 kJ
78. Which of the following metals is the easiest to forge?
    (a) aluminium   (b) mild steel
    (c) cast iron   (d) high-carbon steel
79. The symbol (Fig. 21) represents a
    (a) cell   (b) battery
    (c) earth   (d) switch

Fig. 21

80. Which metal has the greatest tensile strength?
    (a) mild steel   (b) cast iron
    (c) copper   (d) aluminium
81. Brass is an alloy of
    (a) iron and carbon
    (b) aluminium and copper
    (c) tin and copper
    (d) zinc and copper
82. On a drawing a thread is marked M 30 × 2. This means that this is a metric thread with a
    (a) fine pitch of 30 mm
    (b) fine pitch of 2 mm
    (c) coarse pitch of 30 mm
    (d) coarse pitch of 2 mm
83. The flatness of a surface is checked by using a
    (a) dial indicator   (b) try-square
    (c) surface plate   (d) straightedge

84. Fig. 22 shows a hole and shaft combination. The kind of fit obtained is called a
    (a) transition fit
    (b) clearance fit
    (c) press fit
    (d) force fit

Fig. 22

85. Fig. 23 shows a lathe tool. The angle marked A is called the
    (a) clearance angle
    (b) rake angle
    (c) point angle
    (d) wedge angle

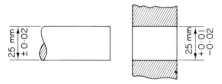

Fig. 23

86. When a horizontal surface is to be produced by milling, which cutter is most suitable?
    (a) side and face cutter
    (b) slab mill
    (c) slitting saw
    (d) slotting cutter

87. When a bar is to be increased in length by reducing the width and thickness of the bar the forging operation is called
    (a) upsetting
    (b) jumping up
    (c) drawing down
    (d) swaging

88. One of the following is flux used when brazing
    (a) zinc chloride (killed spirits)
    (b) resin
    (c) tallow
    (d) borax

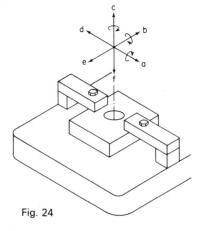

Fig. 24

89. Fig. 24 shows the clamping used whilst drilling a hole. The work is prevented from moving in any of the directions marked a, b, c, d, e and f by either friction or a positive restraint. Which of the following movements is a restrained by friction?
    (a) direction a
    (b) direction b
    (c) direction c
    (d) direction f

90. A body has a mass of 100 kg. Its weight is
    (a) 100 kg
    (b) 981 kg
    (c) 981 N
    (d) cannot be found

91. Fig. 25 shows a plate with a hole cut in it. Which diagram shows the centre-of-gravity of the plate in the correct position?

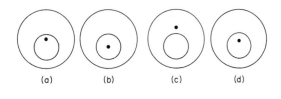

(a)   (b)   (c)   (d)

92. Fig. 26 shows a plate clamp. The force in the bolt is
    (a) 333 N
    (b) 500 N
    (c) 1000 N
    (d) 1500 N

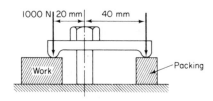

Fig. 26

93. In flame cutting using an oxy-acetylene flame the metal is removed by
    (a) reduction
    (b) oxidation
    (c) combustion
    (d) corrosion

94. The milling operation shown in Fig. 27 is called
    (a) end milling
    (b) gang milling
    (c) upcut milling
    (d) downcut milling

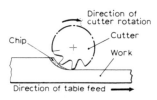

Fig. 27

95. Fig. 28 shows some methods of making a joint by using an adhesive. Which joint is unsatisfactory?

Fig. 28

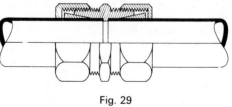

Fig. 29

96. The pipe joint shown in Fig. 29 is called a
    (a) manipulative joint
    (b) flanged joint
    (c) capillary-soldered joint
    (d) non-manipulative joint

Answers to exercises

Exercise 1.1

1. 12 2. 60 3. 12 4. 60
5. 40 6. 24 7. 100 8. 160
9. $\frac{1}{2}, \frac{7}{12}, \frac{2}{3}, \frac{5}{6}$ 10. $\frac{3}{4}, \frac{6}{7}, \frac{7}{8}, \frac{9}{10}$ 11. $\frac{11}{20}, \frac{3}{5}, \frac{7}{10}, \frac{13}{16}$ 12. $\frac{5}{8}, \frac{13}{20}, \frac{2}{3}, \frac{3}{4}$

Exercise 1.2

1. $2\frac{1}{2}$ 2. $2\frac{2}{3}$ 3. $3\frac{1}{7}$ 4. $1\frac{1}{10}$
5. $2\frac{3}{8}$ 6. $1\frac{1}{4}$ 7. $2\frac{2}{3}$ 8. $2\frac{1}{20}$
9. $\frac{21}{10}$ 10. $\frac{13}{4}$ 11. $\frac{39}{32}$ 12. $\frac{131}{64}$
13. $\frac{23}{7}$ 14. $\frac{87}{20}$ 15. $\frac{26}{3}$ 16. $\frac{59}{8}$

Exercise 1.3

1. $\frac{5}{6}$ 2. $1\frac{3}{10}$ 3. $1\frac{1}{8}$ 4. $\frac{9}{20}$
5. $1\frac{7}{8}$ 6. $\frac{401}{1000}$ 7. $3\frac{11}{16}$ 8. $10\frac{1}{15}$
9. $5\frac{23}{64}$ 10. 10 11. $11\frac{3}{16}$ 12. $9\frac{7}{15}$

Exercise 1.4

1. $\frac{1}{6}$ 2. $\frac{4}{21}$ 3. $\frac{1}{6}$ 4. $\frac{2}{5}$
5. $\frac{1}{24}$ 6. $2\frac{1}{4}$ 7. $1\frac{1}{8}$ 8. $2\frac{3}{5}$
9. $1\frac{27}{40}$ 10. $\frac{41}{80}$ 11. $1\frac{49}{160}$

Exercise 1.5

1. $2\frac{3}{8}$ 2. $1\frac{7}{20}$ 3. $7\frac{7}{8}$ 4. $\frac{7}{12}$
5. $\frac{15}{16}$ 6. $\frac{9}{20}$ 7. $\frac{41}{80}$ 8. $3\frac{5}{64}$
9. $1\frac{71}{80}$ 10. $2\frac{39}{100}$

Exercise 1.6

1. $1\frac{1}{3}$ 2. 4 3. $\frac{7}{16}$ 4. $1\frac{1}{2}$
5. $\frac{3}{14}$ 6. $\frac{1}{24}$ 7. 4 8. $6\frac{3}{4}$
9. $\frac{3}{4}$ 10. $8\frac{1}{4}$

Exercise 1.7

1. $1\frac{11}{25}$ 2. 4 3. $\frac{2}{3}$ 4. $1\frac{1}{3}$
5. $1\frac{1}{2}$ 6. $\frac{2}{3}$ 7. $3\frac{1}{2}$ 8. $1\frac{1}{5}$
9. $\frac{25}{26}$ 10. $\frac{7}{15}$

Exercise 1.8

1. 0·3 2. 0·57 3. 0·243 4. 0·008
5. 0·03 6. 0·017 7. 8·05 8. 4·0205
9. 50·004 10. 0·1405 11. $\frac{2}{10}$ 12. $3\frac{1}{10}$
13. $2\frac{46}{100}$ 14. $423\frac{15}{100}$ 15. $\frac{3}{1000}$ 16. $\frac{25}{1000}$
17. $300\frac{26}{1000}$ 18. $42\frac{2385}{10000}$ 19. $\frac{329}{100000}$ 20. $\frac{1}{10000}$

Exercise 1.9

1. 2	2. 4·375	3. 12·29	4. 49·47
5. 46·789	6. 9·17	7. 15·021	8. 0·373
9. 0·822	10. 1·875		

Exercise 1.10

1. 31, 310, 3100	2. 13·2, 132, 1320	3. 0·81, 8·1, 81	4. 2·5, 25, 250
5. 13·85	6. 2700	7. 1700·5	8. 3124
9. 7	10. 513·2		

Exercise 1.11

1. 0·25, 0·025, 0·0025
2. 5·9163, 0·591 63, 0·059 163
3. 0·005, 0·0005, 0·000 05
4. 51·03, 5·103, 0·5103
5. 0·052
6. 0·001 05
7. 0·7803
8. 0·003
9. 0·000 096
10. 0·072 561

Exercise 1.12

1. 697·2992	2. 0·035 073	3. 0·1875	4. 5·276 25
5. 0·000 713			

Exercise 1.13

1. 1·31	2. 0·016	3. 189·74	4. 4·1066
5. 43·2			

Exercise 1.14

1. (a) 18·8657   (b) 18·87   (c) 19
2. (a) 0·008 269   (b) 0·008 27   (c) 0·0083
3. (a) 4·9685   (b) 4·97   (c) 5
4. 49
5. 38·80
6. (a) 40 587 000   (b) 41 000 000
7. (a) 3·1500   (b) 3·150   (c) 3·1
8. 12·20

Exercise 1.15

1. 0·083	2. 0·016	3. 8·578	4. 2·533
5. 0·127	6. 14·557	7. 2·828	8. 163.
9. 0·140	10. 455·111		

Exercise 1.16

1. 0·25	2. 0·75	3. 0·375	4. 0·6875
5. 0·5	6. 0·6667	7. 0·6563	8. 0·6094
9. 1·8333	10. 2·4375		

Exercise 1.17

1. (a) 7316 mm   (b) 731·6 cm
2. (a) 521·6 cm   (b) 5·216 m
3. 804 mm
4. 95·82 cm
5. 8480 mm

Exercise 1.18

1. 498 mm
2. 35·51 mm
3. 1·52 mm
4. A = 8·25 mm; B = 121·98 mm; C = 176·48mm
5. A = 10·84 mm; B = 18·69 mm
6. 1·84 mm
7. A = 5·4 mm; B = 5·4 mm
8. 8·38 mm
9. 35·53 mm
10. 18·355 mm
11. 15
12. 2 cuts; 0·235 mm
13. (a) 10·8 mm   (b) 122
14. 49·95 mm
15. 108 mm
16. 76
17. 37 mm

Exercise 1.19

1. 115	2. 56·83 mm	3. 731 °C	4. 27 m/min
5. 33·405 mm	6. 12·15 kg	7. 3·1625 kg	8. $552\frac{2}{3}$ litre
9. 66·7	10. 5500 hours; 687·5 hours		

Exercise 1.20

1. 2:3	2. 5:8	3. 9:7	4. 5:2
5. $\frac{4}{5}$	6. $2\frac{2}{3}$	7. $\frac{2}{9}$	8. $\frac{2}{3}$
9. 3	10. $3\frac{1}{3}$	11. $\frac{2}{3}$	12. $\frac{3}{5}$

Exercise 1.21

1. 600 mm and 400 mm	2. 20 mm, 40 mm and 100 mm
3. 4·8 kg; 3·2 kg	4. 90 mm, 120 mm and 150 mm
5. 9·8 kg, 2·8 kg and 1·4 kg	6. 292·5 g and 184·5 g
7. 45 mm, 75 mm and 90 mm	8. 30 mm and 45 mm
9. 200, 230 and 250	10. 0·1 kg, 0·2 kg and 0·95 kg

Exercise 1.22

1. £40	2. £1·5	3. 47·5 kg	4. 74 mm; 45 rev
5. 15 mm	6. 1 in 50	7. 130·3 mm	8. 10 days
9. 40 rev/min	10. 160 rev/min		

Exercise 1.23

1. 50%	2. 63%	3. 81·3%	4. 66·7%
5. 72·35%	6. $\frac{2}{5}$	7. $\frac{13}{20}$	8. $\frac{241}{500}$
9. $\frac{3}{50}$	10. $\frac{3}{1000}$		

Exercise 1.24

1. (a) 9 (b) 21·525	2. (a) 16% (b) 16·6% (c) 37·5%
3. 30	4. $73\frac{1}{3}$%, $23\frac{1}{3}$%, $3\frac{1}{3}$%
5. 27 kg and 18 kg	6. 150 kg, 200 kg
7. 30	8. 40 000 kg
9. £451·25	10. 0·9

Exercise 1.25

1.	(a) 6·250	(b) 17·31	(c) 13·12
	(d) 60·56	(e) 26·69	(f) 1444
	(g) 2470	(h) 25 670	(i) 4 769 000
	(j) 263·4	(k) 0·000 625	(l) 0·069 17
	(m) 0·000 014 57	(n) 0·1535	(o) 0·004 865
2.	(a) 1·581	(b) 2·040	(c) 1·903
	(d) 2·790	(e) 2·273	(f) 6·164
	(g) 7·050	(h) 12·66	(i) 46·73
	(j) 4·029	(k) 0·1581	(l) 0·5128
	(m) 0·061 79	(n) 0·6259	(o) 0·2641
3.	(a) 0·4000	(b) 0·2404	(c) 0·2760
	(d) 0·1285	(e) 0·1935	(f) 0·027 78
	(g) 0·020 12	(h) 0·006 242	(i) 0·000 457 9
	(j) 0·061 62	(k) 40·00	(l) 38·02
	(m) 263·0	(n) 2·553	(o) 14·34
4.	(a) 5·801	(b) 29·34	(c) 9·879
	(d) 24·06	(e) 0·4156	(f) 0·6911
	(g) 16·06	(h) 0·044 24	(i) 0·4807
	(j) 0·4047		

Exercise 1.26

1. 15	2. −25	3. −17	4. 13
5. −4	6. −8	7. −17	8. 4
9. 6	10. −18	11. −16	12. −1
13. 4	14. −4	15. −13	16. 3
17. 11	18. −4	19. −11	20. 7
21. 20	22. −20	23. −20	24. 20
25. 9	26. 64	27. −18	28. 2
29. 2	30. −2	31. −2	32. 2
33. 2	34. −1	35. −1	36. −6

Exercise 1.27

1. 0·5051	2. 0·6628	3. 0·3636	4. 0·9170
5. 0·9952	6. 0·4991	7. 0·9143	8. 0·8518
9. 0·2732	10. 0·6626		

Exercise 1.28

1. (a) 0·9191	(b) 1·9191	(c) 2·9191	(d) 3·9191
2. (a) 0·6243	(b) 1·6243	(c) 2·6243	(d) 3·6243
3. (a) 1·9134	(b) 2·9134	(c) 3·9134	(d) 4·9134
4. (a) 2·8535	(b) 3·8535	(c) 4·8535	(d) 5·8535
5. (a) 0·9337	(b) 1·6278	(c) 2·9104	(d) 0·2092

Exercise 1.29

1. 6·457	2. 57·54	3. 107·4	4. 1466
5. 23·02	6. 3·987	7. 1442	8. 46 410
9. 10·91	10. 122 800		

Exercise 1.30

1. 16·32	2. 464·7	3. 929·8	4. 91 880
5. 2558	6. 2000	7. 3·997	8. 13·81
9. 3·120	10. 2·569		

Exercise 1.31

1. (a) 1·4852	(b) 0·4852	(c) $\bar{1}$·4852	(d) $\bar{3}$·4852
2. (a) 0·8029	(b) $\bar{1}$·8029	(c) $\bar{2}$·8029	(d) $\bar{3}$·8029
3. (a) $\bar{1}$·8670	(b) $\bar{2}$·3345	(c) $\bar{4}$·6827	(d) $\bar{1}$·7758
4. (a) $\bar{2}$·2608	(b) $\bar{3}$·9921	(c) $\bar{5}$·7482	(d) $\bar{2}$·8880

Exercise 1.32

1. 0·3947	2. 0·039 47	3. 0·003 947	4. 0·004 035
5. 0·014 74	6. 0·000 129 0	7. 0·3166	8. 0·008 337
9. 0·010 16	10. 0·000 041 80		

Exercise 1.33

1. $\bar{1}$	2. 2	3. $\bar{5}$	4. $\bar{4}$
5. $\bar{7}$	6. 0	7. 1	8. 2
9. $\bar{4}$	10. $\bar{1}$	11. 1·7	12. $\bar{4}$·9
13. 2·5	14. $\bar{3}$·2	15. $\bar{4}$·3	16. 1·7
17. 0·5	18. 0·1	19. 1·3	20. $\bar{1}$·3
21. 2	22. $\bar{1}$	23. 1	24. 2
25. 0	26. 3	27. $\bar{2}$	28. 5
29. $\bar{1}$	30. $\bar{1}$	31. $\bar{4}$	32. $\bar{3}$
33. 0·7	34. $\bar{3}$·1	35. 0·3	36. 1·2
37. 2·2	38. $\bar{1}$·9	39. $\bar{3}$·7	40. $\bar{1}$·7
41. $\bar{4}$·7	42. 0·8	43. 2·9	44. 4·5

Exercise 1.34
1. 1·104
2. 3·583
3. 0·4277
4. 0·018 73
5. 24·13
6. 59·44
7. 0·001 462
8. 28 930
9. 1·697
10. 0·001 407
11. 0·1477
12. 95·23
13. 0·1710
14. 11·43
15. 11·50

Exercise 1.35
See answers in Exercises 1.30 and 1.34.

Exercise 1.36
1. $7x$
2. $4x-3$
3. $5x+6$
4. $\frac{x+y}{z}$
5. $\frac{1}{2}x$
6. $8xyz$
7. $\frac{xy}{z}$
8. $3x-4y$

Exercise 1.37
1. 6
2. 1
3. 4
4. 8
5. 15
6. 2
7. 18
8. 6
9. 48
10. 10
11. 13
12. 14
13. 20
14. $\frac{1}{2}$
15. $\frac{1}{2}$
16. $\frac{1}{3}$
17. $\frac{1}{4}$
18. $1\frac{1}{2}$
19. $1\frac{2}{3}$
20. $3\frac{1}{2}$

Exercise 1.38
1. 8
2. 81
3. 18
4. 24
5. 188
6. 101
7. 20
8. 756
9. 3
10. $21\frac{1}{3}$

Exercise 1.39
1. 3
2. 2
3. 4
4. 8
5. 5
6. 5
7. 3
8. 4
9. 3
10. 10
11. 21
12. 18
13. 8
14. 4
15. 6
16. 2
17. 7
18. 2
19. 2
20. $\frac{1}{2}$
21. 6
22. 5
23. $7\frac{1}{2}$
24. $\frac{1}{3}$

Exercise 1.40
1. A = 1·5 mm
2. B = 14·7 mm
3. X = 15·55 mr ı
4. 400
5. 50 mm
6. 13

Exercise 1.41
1. 5
2. 16
3. 250
4. 0·05
5. 2000
6. 190
7. 440
8. 616
9. 17·6
10. 1·5

Exercise 1.42
1. (a) 16 mm²
   (b) 625 cm²
   (c) 64 m²
2. (a) 40 m²
   (b) 176 cm²
   (c) 875 000 mm²
3. (a) 332 mm²
   (b) 900 mm²
   (c) 2400 mm²
4. (a) 300 mm²
   (b) 3 cm²
5. 8000
6. 12 500 mm²
7. 1·026 kg
8. 693 mm²
9. 347 mm²
10. 3850 mm²; 220 mm
11. 616 mm²; 88 mm
12. 571·3 mm²
13. 440 m
14. 13 200 mm
15. (a) 7·07 mm
    (b) 14·14 mm
    (c) 28·28 mm
16. (a) 46·28 mm
    (b) 95·12 mm
    (c) 72·38 mm

Exercise 1.43
1. 140 000 mm³
2. 600 000 mm³
3. 7 700 000 mm³
4. 31 420 mm³
5. 187 367 mm³
6. 12 640 mm³
7. 9000 litre
8. 603 litre
9. 1200 cm³
10. 0·003 176 m³
11. 1188 cm²
12. 1413 cm²

Exercise 1.44

1. 9·746 kg
2. 16·02 kg
3. 20·15 kg
4. 85·68 kg
5. 3532 kg
6. 2·325 kg
7. 0·41 kg
8. (a) 8 km
   (b) 15 Mg
   (c) 3·8 Mm
   (d) 1·891 Gg
   (e) 7 mm
   (f) 28 mg
   (g) 360 mm
   (h) 3·6 mA

Exercise 1.45

1. 12·6 m/min
2. 31·4 m/min
3. 265 rev/min
4. 68 rev/min
5. 1900 rev/min
6. 540 rev/min
7. 10 min
8. 0·39 min
9. 159 rev/min
10. 159 rev/min
11. 113 m/min
12. 4 min
13. 24 min
14. (a) 120 rev/min
    (b) 168 mm/min
    (c) 1·2 min
15. 110 rev/min

Exercise 1.46

2. 10·2 mm
3. 200 rev/min
4. 89%
5. 6 kg; 3·2 kg
6. 830 rev/min
7. 720 W

Exercise 1.47

2. (a) angle A = 59°; angle B = 175°
   (b) angle A = 31°; angle B = 47°
   (c) angle A = 54°; angle B = 7°
3. (a) 18°
   (b) 45°
   (c) 90°
   (d) 120°
   (e) 108°
4. (a) 180°
   (b) 54°
   (c) 30°
5. 36°
6. (a) 55°
   (b) 33° 34′
   (c) 79° 25′
   (d) 59° 1′
   (e) 107° 15′
7. (a) 18° 7′ 45″
   (b) 33° 18′ 22″
   (c) 96° 32′ 28″
   (d) 86° 2′ 11″
8. (a) 7° 5′
   (b) 6° 18′
   (c) 3° 13′ 53″
10. (a) 20°
    (b) 22°
    (c) 11°
11. 36·7°
12. (a) 27·31°
    (b) 49·46°
    (c) 64·83°
    (d) 11·29°
13. 18° 42′
14. (a) 75° 48′
    (b) 3° 33′ 36″
    (c) 37° 49′ 30″

Exercise 1.48

1. (a) 81°
   (b) 147°
   (c) 53°
2. (a) 70° each
   (b) 57°; 66°
   (c) 72°; 36°
3. (a) 18°
   (b) 9°
   (c) 2·5 cm
4. (a) 69°
   (b) 21°
   (c) 4·5 cm
5. (a) 50 mm
   (b) 9·17 cm
   (c) 130 mm
6. (a) 4·47 cm
   (b) 70·33 mm
   (c) 279 mm
7. (a) 2·65 cm
   (b) 3·05 cm
   (c) 74·2 mm
   (d) 82·6 mm
8. 42·43 mm
9. 9·43 cm
10. 35·36 mm
11. 17·7 mm
12. 6·02 cm
13. 44·72 mm
14. 405·6 mm

Exercise 2.1

1. 1000	2. 1000	3. 4905 N	4. 882·9 N
5. 19 620 N	15. 40 kN/m²	16. 125 kN/m²	19. 0·001

20. 500 kN/m²; 0·000 033
23. (a) 18 N acting to the right   (b) 5 N acting to the left
     (c) 480 N acting downwards   (d) 200 N acting downwards
24. (a) 53 N acting at 19° to the horizontal   (b) 129 N acting at 18° to the horizontal
     (c) 166 N acting at 20° to the horizontal   (d) 600 N acting at 52° to the horizontal
     (e) 538 N acting at 68° to the horizontal   (f) 1120 N acting at 63° to the horizontal
25. 608 N acting at 81° to the horizontal   26. 866 N
27. 500 N   28. (d)

Exercise 2.2

1. 2400 N	2. 100 N	3. 20 N
4. 125 N	5. 100 N m	6. 8 N m
7. 200 N	8. 4800 N	9. 400 N; 1200 N

10. $Q = 4, R = 12$; $P = 16, R = 20$; $R = 14, y = 7·5$; $R = 14, y = 5·5$; $R = 17·5, x = 9$
11. 200 N   12. 20 cm from the support

Exercise 2.3

5. 50·015 cm	6. 79·30 cm	7. 50·0056 mm	8. 182·36 mm

Exercise 2.4

4. 5852 kJ	5. 1840 kJ	6. 29·25 kJ	7. 0·39
8. 9200 kJ			

Exercise 2.6

1. 800 J	2. 10 kJ	3. 9·81 kJ	4. 23·5 kJ
5. 14 kJ	6. 1·76 kJ	7. 50 W	8. 200 W
9. 750 W	10. 785 W	11. 0·45 kW	12. 0·9 kW

Exercise 2.10

9. 100 coulombs	10. 75 coulombs	11. 2·4 kW	12. 12 A
13. 4 kW h	14. 0·8 kW h	15. £2·80	18. 0·5 A
19. 8 A	20. 250 V	21. 160 V	22. 30 Ω
23. 4 Ω	24. 60 Ω	25. 4 A	26. 1 A
27. 4 V and 8 V	28. 1·2 Ω	29. 9 A	30. 6 A and 3 A

31. 24 V   32. 1 and 3 are ammeters; 2 is a voltmeter
33. 1 and 3 are ammeters; 2 is a voltmeter   34. 20 A
35. 1 reads 4 V; 2 reads 8 V   36. 1 reads 2 A; 2 reads 8 V

Index